Quality Control of Packaging Materials in the Pharmaceutical Industry

Packaging and Converting Technology

A Series of Reference Books

edited by

Harold A. Hughes
Director, School of Packaging
Michigan State University
East Lansing, Michigan

1. Packaging: Specifications, Purchasing, and Quality Control, Third Edition, Revised and Expanded, *by Edmund A. Leonard*

2. Medical Device Packaging Handbook, *edited by Joseph D. O'Brien*

3. Integrated Packaging Systems for Transportation and Distribution, *by Charles W. Ebeling*

4. Quality Control of Packaging Materials in the Pharmaceutical Industry, *by Kenneth Harburn*

Other Volumes in Preparation

Quality Control of Packaging Materials in the Pharmaceutical Industry

Kenneth Harburn

Glaxo Manufacturing Services Limited
Barnard Castle, England

CRC Press
Taylor & Francis Group
Boca Raton London New York

CRC Press is an imprint of the
Taylor & Francis Group, an **informa** business

CRC Press
Taylor & Francis Group
6000 Broken Sound Parkway NW, Suite 300
Boca Raton, FL 33487-2742

First issued in paperback 2019

© 2008 by Taylor & Francis Group, LLC
CRC Press is an imprint of Taylor & Francis Group, an Informa business

No claim to original U.S. Government works

ISBN-13: 978-0-8247-8243-6 (hbk)
ISBN-13: 978-0-367-40310-2 (pbk)

A CIP record for this book is available from the British Library.

Library of Congress Cataloging-in-Publication Data available on application

Visit the Taylor & Francis Web site at
http://www.taylorandfrancis.com

and the CRC Press Web site at
http://www.crcpress.com

Preface

To produce a good quality packaged product, all areas requiring quality control must be identified and the correct quality control system operated. With packaging materials, quality control starts at the design stage and continues through to the customer using the packaged product. This volume covers quality control aspects through every stage. Once the quality is right, user safety is assured, product filling and packaging will run efficiently, and the customer will receive the order on time. Quality and efficiency are closely related and hence good quality control must be cost effective.

Although it is easy to list the systems required to control quality, it can be difficult to actually set up a working system. When theory is put into practice, unforeseen, quite often seemingly insurmountable problems come to light. Unless all the difficulties are identified and resolved, then what managers think is actually working quite often is not.

This book is intended to be both an aid to setting up systems and a guide on how to approach problems. Once the mind is focused on the way to think through an idea, then setting up such systems as in-process control, quality auditing, and specifications will work first time.

The volume should be useful to component and packaging machine manufacturers as well as the pharmaceutical industry, showing how they need to work together to minimize problems and give the highest possible assurance of product quality. Although most of the discussion is specifically aimed at the pharmaceutical industry, the same principles operate in successful food and cosmetic industries.

I would like to thank Mrs. Margaret Rhodes, Mr. Peter Williams, Mr. Derek Le Mare, Dr. Roger Horsgood, Dr. Roy Tranter, and Mr. Andrew Rixon for contributing technical comments. Thanks also to Mr. Ian Rickard for preparing the computer-generated figures, and a special thanks to Miss Angela Jeremy for having the patience to interpret my writing while typing the manuscript.

Kenneth Harburn

Contents

Preface   *iii*

1 Packaging Design and Specifications   **1**
   I. Introduction   1
   II. Packaging Design   1
   III. Component Specifications   19

2 Supplier Quality Auditing   **45**
   I. Introduction   45
   II. Quality Auditor   45
   III. Auditing   47
   IV. Problems Encountered by Auditors   57
   V. General Comments Concerning Suppliers   60

3 **Quality Control at the Supplier's Premises**   **61**
    I. Introduction   61
    II. GMP Deficiencies   62
    III. Component Specification   81
    IV. Communication of Quality Problems   81
    V. Summary   83

4 **The Packaging Material Quality Control**
   Laboratory on the Pharmaceutical Premises   **85**
    I. Introduction   85
    II. Laboratory Consideration   86
    III. Receipt, Sampling, Testing, and Sentencing of
       Packaging Materials   90
    IV. Documentation   94
    V. Investigating Component Problems Occurring
       Within the Pharmaceutical Premises   95
    VI. Sterilization of Primary Packaging
       Components by Gamma Irradiation   100

5 **Packaging and Filling Equipment**   **107**
    I. Introduction   107
    II. Equipment Specification   108
    III. Equipment Purchase   116
    IV. Installation of New Equipment   120
    V. Commissioning of New Equipment   125
    VI. Validation   127
    VII. Equipment and Component Manufacturer Interface 131

6 **Pharmaceutical Packaging**   **133**
    I. Introduction   133
    II. Planning the Work   135
    III. Component Storage and Prepreparation   136
    IV. Preparation Area   137
    V. Sterile Area   143
    VI. Filled Sterile Product Receiving Area   149
    VII. Collation Area   155

VIII. Packaging Area   156
IX. Staff Training   165

7 Customer Complaints   **167**
I. Introduction   167
II. Complaint Procedure   167

Index   *175*

Quality Control of
Packaging Materials in the
Pharmaceutical Industry

1

Packaging Design and Specifications

I. INTRODUCTION

Quality control of a packaging component starts at the design stage. All aspects of a pack development that may give rise to quality problems must be identified and minimized by good design. Identifying mistakes when a pack is in use can have serious consequences, which at worst could result in a product recall and at least in excessive rejects or low production rates.

II. PACKAGING DESIGN

Considering the implications of a badly designed pack, one suitably experienced person should be assigned the responsibility of designing a pack. This packaging design coordinator must be fully aware of the involvement required by marketing, quality assurance, and production.

There are several important considerations to bear in mind from the start.

Product quality must be maintained until the customer uses the product.

Customers must be able to easily access and use the product without harming themselves or contaminating the product, yet the pack should be tamper evident and child proof.

The component manufacturers must be capable of making the components to the required specifications.

The pharmaceutical equipment must be capable of handling the components and maintaining product quality and production efficiency.

The first consideration is the product, therefore, a brief is required from the product development staff stating the product details, including the type of protection required from the environment, for instance, is the product susceptible to degradation in the presence of moisture, light, oxygen, etc.?

Marketing requirements need specifying, such as the pack presentation, for example, whether a foil or container pack for a tablet product is required and the size or sizes of packs to be marketed. This information should be based on market research and an assessment of the market complaints for previous similar products (where possible).

With regard to market research, for example, it is vital that marketing (and the markets) have an input into the method of product administration at the development stage. A customer finding a product difficult to administer may prefer to use a similar competitor's product that is easier to administer.

When all the data is available, the pack designing can start. This information should be obtained as early as possible during product development to ensure that sufficient time is available for designing the pack requirements. This will minimize the delay time for launching the product.

A. Component Shape and Dimensions

Standardizing both component shape and size should be the policy. There may be product-specific design requirements in some instances, but there are many components that can be standardized, such as ampules, vials, cartons, labels, and leaflets. Rubber plugs and plastic bottles can be standardized with respect to shape and size, varying only in the material of construction. There will be a variety of sizes of components depending on the dosage, but again the same shape could be used but with different dimensions. There are several advantages to standardizing components. These are

1. Minimizing the packaging validation trials required.
2. Enabling larger quantities of unprinted components to be ordered at a time, which may make them cheaper to buy.
3. Reducing the number of packaging machine changeovers for different types of packs, i.e., fewer change parts and resetting of equipment. This can reduce the number of packaging rejects and improve packaging efficiency.
4. Shortening the lead time for supplying a customer order by allowing a standard stock of filled product to be allocated, requiring only the final packaging to the specific customer requirements, e.g., labelling and cartoning for a foreign customer.
5. Allowing more utilization of the packaging equipment available and also minimizing the variety of packaging equipment required. This can have further advantages in that fewer spare parts will be needed to be stored and less staff will have to be trained in equipment usage, maintenance, and repair.
6. Fewer component drawings will be required, reducing the administrative work in preparing, updating, and distributing drawings.
7. Reducing inventory—the cash released can earn money at money market rates.

When the components required for a pack have been identified, it is not a good idea for the packaging design coordinator to decide the pack design in isolation. The component manufacturers should be consulted, as they are the experts in the best design considerations to maximize the output and minimize problems within the component manufacturer's production. They will also be able to draw on the experience of their other customers. Requesting a ''one-off'' pack design may create component manufacturing problems, and hence an increased possibility of defective components being supplied to the pharmaceutical company. Nonstandard production components will probably be more expensive to buy.

B. Marketing Involvement

Consultation with marketing staff should take place when a basic, practical design has been developed. They may have ideas for attractive component designs that may help sell over-the-counter pharmaceutical products, although the prime consideration must be the product safety and security. The possible sales advantages against product packaging capabilities must be considered. Some design concepts may be expensive to produce and could cause product quality problems.

Once a final pack design has been agreed, the emphasis should be on minimizing the variations of packs for different markets (except for the inevitable artwork variations for foreign countries).

One possible method of minimizing variations is to provide a product prospectus to the sales staff. This should show a picture of each standard product pack plus the various components included in the pack. A potential customer can then be shown the packs available. With prescription products, a salesperson cannot carry samples (unless they contain placebos). This can be a potential security risk and should be avoided, therefore, the prospectus is the best alternative. If a customer cannot be shown the types of packs available, it is possible that a nonstandard pack will be requested.

The prospectus must be kept up to date for the system to work effectively, therefore, a record of who has copies is required. Also, a mechanism for the removal of old packs and the introduction of new packs into the prospectus is required.

C. Packaging Validation Trials

When the components have been identified for a particular product, the validation of the packaging operation is required. This is to ensure that a consistent pack quality is obtained at the required packaging rate. Before starting, carefully plan the stages of the validation study. Prepare a validation protocol detailing all the tests required and the standards to be attained. An example could be the foiling and packaging of tablets. Each specific stage of the foiling and packaging operation should be identified and proved to be working correctly. The possible stages of a tablet foiling and packaging operation are shown in Figure 1.1.

When each stage has been identified, determine the method and rate of monitoring for each operation to ensure that the quality is satisfactory.

At the foiling stage, the heat-sealing operation must give a consistent leakproof seal, print a legible batch number and expiry date, and not damage the tablet or the foil, and ensure that each pocket of the foil strip contains a tablet. Therefore, the parameters requiring close monitoring are the appearance of the foil strip, the appearance of the tablets, and the effective sealing of each foil. The quality of the seal can be monitored by a simple dye test. Immerse the sealed foils in a dye (methylene blue) contained in a suitable evacuation vessel fitted with a calibrated gauge. Then pull a vacuum for one minute. Release the vacuum, leave the foils immersed for another minute, wash the foil strips, then carefully open each pocket and look for signs of dye penetration.

The rate of monitoring should be determined by the foiling rate, but must be based on a statistical sampling plan that will ensure that the packaging specification quality can be complied with.

To obtain a consistent, satisfactory seal will require careful monitoring of the heat-sealing roller temperature, pressure, and dwell time. Once the optimum conditions have been determined, the machine must always be operated within these validated limits.

The same approach should be used for the assessment of the rest of the packaging operation.

This monitoring exercise should be performed at both the mini-

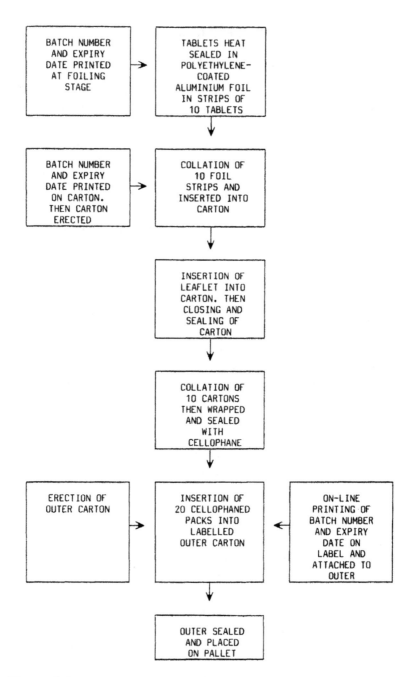

Figure 1.1. Possible stages of a tablet foiling and packaging operation.

mum and maximum packaging rates at which the packaging line is to be operated. Also, at least three batches of product, and preferably different batches of components, should be monitored.

The results obtained must be recorded and a validation report should be prepared that details the packaging equipment used, packaging rates, aspects monitored, methodology, results, and conclusions.

If the same packaging equipment is used for different sizes and shapes of tablets or pack sizes, then the validation exercise should be repeated. Very little extra work will be required for different products, provided the tablet shape has been validated through the packaging stage and the same type of foil is used. Occasionally, with sensitive tablets, additional monitoring may be needed through the foiling stage. For example, extremely hygroscopic tablets may pick up moisture during foiling. This may require a low-humidity foiling area, or localized low-humidity control on the machine may be required and special precautions may be taken during prolonged stoppages such as breakdowns or lunch breaks, when the feed hopper may need to be emptied.

With filled ampules, vials, plastic bottles, and collapsible tubes, the packaging validation work will not need repeating once a particular size has been validated.

The replacement of equipment on a packaging line with updated or modified equipment will require the revalidation of the packaging operation concerned.

D. Material of Construction

The material of construction requires careful consideration, particularly when the product is in contact with the container. It is necessary to ensure that the product does not deteriorate or does not become contaminated as a result of being in contact with the container, or that the product does not affect the integrity of the pack.

Therefore a series of experiments will be required to ensure that the product/pack integrity is maintained. A good example to consider is a plastic bottle containing a product. The bottle formulation details need to be agreed upon with the manufacturer, that is, the complete list and quantities of each ingredient should be reviewed. Also, an assurance that these formulation details are not changed without prior consultation with the pharmaceutical company concerned is required. Included in this must be changes in the supplier of the bottle ingredients. It may be necessary to enter a secrecy agreement with the supplier. Authorities such as the FDA and overseas licensing authorities will require such details before a manufacturing license is issued.

The main reason for the tight control over container manufacture is to ensure that changes are not made that can cause product degradation, particularly if the product is a liquid.

E. Component/Product Validation

Once a formulation has been agreed, the pharmaceutical company has to perform compatibility studies between the product and container to ensure that product degradation does not occur during the product market life. Also, the container has to be capable of protecting the product from the environment. For example, it may need to be light and moisture proof. Product stability studies will identify such requirements. The following two examples show the approach required for a sterile and nonsterile product, respectively. They show how important it is to base the validation work on the product and pack characteristics.

1. Sterile Product/Pack Validation

Suppose a sterile liquid, injectable product was to be marketed in a glass vial with an inert atmosphere (e.g., nitrogen). To start with, the components and product to be used must be fully tested and passed to the relevant specification requirements. Each component must be validated through the washing and sterilization stages. Then the development work required relating to the pack would be as follows.

a. Product and Pack Compatibility. The components must be washed and sterilized through the validated procedure. Then the vials must be filled with sterile product under sterile conditions and terminally sterilized if this is part of the intended production operation. Full quality control procedures must be operated and recorded during the filling and terminal sterilizing processes, then the product and pack compatibility can begin. This involves storage at each of the market temperatures for one year beyond the expected market life of the product. This should involve at least three batches of product, with sufficient samples at each market temperature to enable full chemical testing of the product periodically throughout the storage period. Samples containing the amount of product to be marketed in each pack should be stored in an upright and inverted position. This is to determine the glass vial/product compatibility and the rubber plug/product compatibility. The majority of the sample must be stored in the dark, which is the storage method expected to be used on the market. Stability trials under daylight conditions should be performed to determine the maximum exposure time to daylight.

The storage areas for the compatibility trials should be accurately controlled to the recommended product storage temperatures.

Accelerated stability trials can sometimes prove useful in determining the compatibility of the pack and product, but cannot be relied upon to give an accurate predicted market life. A high temperature, such as 50°C, is usually used, and quite often product or pack deterioration occurs, whereas under normal temperatures problems are not encountered.

Component performance should be monitored during the compatibility trials to ensure that deterioration has not occurred. In particular, the rubber plug may perish, or in the case of a multidose injection, it may fail the self-sealability test or excessive coring may occur. These parameters should be monitored at the same time as the product is checked for stability.

b. Seal Integrity.   Trials to ensure that a sterile seal can be consistently achieved are essential. The components to be used must be checked for the critical dimensions to ensure that they are within the specification limits. Components with the largest dimension variations will ensure a realistic sterility challenge experiment. The components should be individually identified and the dimensions of each component recorded for cross-comparison at the end of the experiment. (Failure to do this may make a full analysis of the experiment difficult, because components such as rubber plugs and aluminum overseals cannot be accurately measured after being used).

The vials should have a quantity of a sterile growth media placed into them. Also the components must be sterilized by the same methods to be used in normal production. The seals of each vial should be examined before the experiment to ensure that there are no defectives, then each vial should be inverted into a tray containing the challenge bacteria. The samples should be cycled through temperature and pressure changes expected on the market for several weeks. Careful cleaning of the vials and examination of the contents for sterility will determine the seal quality. If failures occur, then the components should be carefully examined and correlated to their dimensions to try to determine the cause of failure. It may be that one or more of the critical dimensions have limits that are too wide for the forming of a sterile seal.

One rule that should always be followed during compatibility or pack integrity studies is to examine all samples prior to performing any of the tests, carefully noting product abnormalities (e.g., discoloration) or pack defects. This can sometimes give the answer to unexpected results.

2. Nonsterile Product/Pack Validation

As a comparison, a nonsterile product may still require extensive trials to ensure its suitability for the market.

If we consider a moisture-sensitive tablet packaged in a polypropylene screw-cap bottle, the following validation work will be required.

a. Water Vapor Permeability.   The water vapor permeability of the pack containing the product is required. This is necessary because, although the bottles will comply with the water-vapor permeability test described in the USPXXII (page 1575), permeation through the bottle wall will depend on whether the product has a high or low affinity for water. Therefore, tests for permeation through the bottle wall and seal should be performed with packs containing the full quantity of tablets.

The test should be split into two parts to enable the maximum amount of information to be obtained and hence possibly eliminating the necessity to perform further time-consuming experiments. Prepare sufficient samples to enable full analysis of the product periodically through the proposed market life.

(1) Bottle Wall Permeation. The first experiment will determine the permeation through the bottle wall only. Fill each bottle with tablets and immediately seal over the mouth of each bottle with a polyethylene-coated aluminum foil. Check a sample of the sealed bottles to ensure that a good seal has been obtained. Store the samples under controlled temperature and humidity conditions, that is, the highest expected market storage (or recommended) temperature and the highest expected humidity. If a recommended storage temperature has not been determined for the product, then storage at a range of temperatures from refrigerator, cool, room temperature, and an elevated temperature can be used. Periodic analysis of the product, say, every 3 or 6 months, will determine the market life and storage temperature. It is very important that a sample of the product sealed into the bottle is analyzed immediately after sealing for moisture content and active ingredient using a stability indicating method. These are the starting results for the test.

(2) Bottle and Cap Permeation. The same experiment should be repeated with bottles filled with product and with the cap in place, but without the aluminum seal over the bottle mouth. It is likely that the moisture uptake through the bottle-cap threads will be high compared to the bottle wall (particularly with a hygroscopic product). Therefore, the product analysis should be more frequent, possibly on a daily basis initially. This test may show quickly whether just the capped bottle is sufficient to protect the product from moisture uptake. If not, then it will be necessary to protect the product with an aluminum seal over the mouth of the bottle. This would also be a tamper-evident seal.

The data from this test would determine the life of the product once the seal has been broken. The product would be expected to be used within a few weeks of breaking the seal.

There may be a problem if a pharmacist removes the tablets from the original pack and places them in one of his or her own bottles. If this is likely to cause degradation of the product, then a suitable warning should be stated on the label and leaflet.

b. Light Transmission. This test is to determine the effect of light passing through the bottle wall on the product stability and appearance. The bottle wall thickness can have a significant effect on the results obtained.

Although it is not expected that any product be exposed to direct sunlight for a significant length of time, it is likely that either storage by the pharmacist or customer may result in limited storage of the product in daylight and removal from the carton (if present). Therefore, samples stored in both artificial light and daylight should be checked for product degradation and tablet discoloration.

c. Product/Pack Stability. It is unlikely that a compatibility problem, particularly with a film-coated tablet, will occur, although it is still necessary to check up to the full life of the product. There is a possibility that either the smell or taste of the tablets will be affected. The complete packs should be stored at each of the recommended market storage temperatures and fully analyzed periodically throughout the storage period. The tests should include close examination at the tablet/container contact points, and of smell, and taste.

F. Product Considerations

The product must be carefully considered during packaging design. For example, with tablets, the shape can be very important. The packaging operation can be influenced if the tablet is the wrong shape for the filling/packaging equipment concerned. This is a situation where tablet shape and pack are closely related. Problems that may occur may be poor feed of the tablets at the filling stage, resulting in damaged or missing tablets and a slow filling rate. An outside influence in this area may be the marketing department wanting a particular shape or pack. The advantages and disadvantage of both marketing and packaging must be carefully considered before a final decision is made. There may also be tablet manufacturing aspects to consider.

G. Customer Usability

Customers must be able to open the pack and remove the contents without harming themselves or contaminating the product. For example, a product contained in a glass ampule must be capable of being opened without the ampule breaking and without the customer cutting a finger. Therefore, trials at the design stage are essential to ensure that the glass thickness is correct and that the method of weakening the ampule at the neck is sufficient to enable an easy, clean opening. This will be a combination of constriction diameter, glass thickness, and, with a ring-snap ampule, a combination of the paint/powdered-glass mixture painted and heat-treated into the constriction. The correct combination of all these parameters must be determined and clearly defined in the component specification. Extensive ampule opening trials (by hand) will be necessary to determine the correct parameters. Instructions on how to open the ampule must be contained in the complete pack supplied to a customer.

The same detailed consideration as to how a customer is to access and use the product should be given to each pack at the design stage.

H. Print Requirements

There are several aspects to consider with all printed components. It is **very important** that the artwork is correct (an incorrectly printed product strength could be disastrous). Also, the legal requirements for the country for which the product is intended must be checked and complied with. The artwork must be checked by a competent person, who is fully aware of the labelling regulations, product details, and the implications of any mistake missed.

The print color should be chosen carefully. Try to ensure that a different print color is used for each product, but in particular, a different color for each strength of a product. This will reduce the risk of a rogue item being used undetected. The print must be large enough for the pharmacist and patient to read, clearly showing the product name, strength of active ingredient, quantity, batch number, storage requirements, product license number, and instructions.

Instructions on labels, leaflets, or cartons must be clear and concise, and where applicable diagrams should be used; for example, instructions on how to open an ampule. It is worth remembering that "a picture is worth a thousand words," and is much easier to follow than written words. A little bit of thought can reduce the number of complaints by a doctor or patient because of misleading instructions on a pack.

When the pharmacist dispenses a product, a label is usually placed on the container or carton, detailing the dosage rate and any special instructions. Quite often this label covers very important information or instructions. Therefore, allowing a blank space on the pack for the pharmacist's label is worth considering when designing the artwork.

I. Regulatory Requirements

Prior to the sale of a product, the regulatory requirements of the country in which the product is to be sold must be met. It is not intended that the administrative aspects relating to submissions will be covered, only the packaging details that need to be submitted to the authorities.

The whole issue of regulatory requirements can be difficult to understand in deciding what is required for new products, how it is to be presented, and what changes to processes require notification to the authorities. In the United States, the Food and Drug Administration (FDA) has to be supplied with sufficient data to be satisfied that the drug is both effective and safe prior to the release of the drug on the market. The submission must comply with the code of Federal Regulations (21CFR part 314) before clearance is given. If insufficient data is supplied, delays will occur in the release of the drug. Too much data may lead to the pharmaceutical company having to submit a great deal of data on minor changes in the annual review to the FDA, since all data submitted concerning chemistry, manufacturing, and controls have to be updated. Supply sufficient data for the FDA to thoroughly review the process without being overly specific. The main requirements are as follows:

1. *New drug application (NDA)*
    This must be submitted and approved prior to the sale of the product on the market.
2. *Supplemental New Drug Application Type I*
    This must be submitted when major process changes are required. The supplemental NDA must be approved before the change takes place.

3. *Supplemental New Drug Application Type II*
    This is a change important enough to require a supplement but does not require FDA approval prior to implementation.
4. *Annual NDA Reports*
    On the anniversary of the approval date of an NDA, a report must be submitted that details the changes that have taken place over the year that do not require a supplemental application.
5. *Drug Master File (DMF)*
    There are five types of DMF. A DMF is a submission to the FDA that may provide detailed information about facilities, processes, or articles used in the manufacturing, processing, packaging, and storing of one or more pharmaceutical products for human use. It is not legally required that a pharmaceutical company submit a DMF.

Consider each of the five regulatory requirements in more detail:

1. New Drug Application

The details that need to be included in this documentation are as follows:

1. Formulation details for all primary components.
2. Primary component drawings indicating the critical dimensions.
3. Primary and secondary component artwork, for example, labels, leaflets, cartons, and printed primary components.
4. Primary component quality standards and the testing methods used, this must include compliance with USP and CFR requirements.
5. Primary component compatibility data, for example, the bottle and cap.
6. Product/primary component compatibility data. Sufficient information must be supplied to confirm compliance with the product specification for the market life of the product.
7. Details of the filling and packaging operations, including the in-process checks operated. (There is no need to specify the exact machinery.)

8. Filling and packaging facility details.
9. A list of the primary component manufacturers, including the polymer manufacturers for plastic components.

2. Supplemental New Drug Application Type I

The type of major changes requiring FDA approval prior to implementation are as follows:

1. Relaxing of primary component specification limits. Supporting data must be supplied that confirm that the change will not affect the quality or performance of the component.
2. Deletion of a specification or limit. Supporting data must be supplied that confirm that the change will not affect the quality or performance of the component.
3. The use of a different facility or establishment for filling or packaging the product, where the facility or establishment differs materially from the former facility or establishment or has not received a satisfactory CGMP (FDA) inspection within the previous 2 years covering the filling or packaging process.
4. Changing a primary component specification relating to the critical dimensions ensuring an effective closure system or changing the container size (except for sealed dosage forms). Supporting data must be supplied that confirm that the change will not affect the quality of performance of the component.
5. Changing a regulatory analytical method for the testing of a primary component.
6. Any change in the artwork text, except those listed in the sections 3.2 and 3.3 for type II supplemental changes.

3. Supplemental New Drug Applications Type II

The types of changes that can be implemented prior to FDA approval of the supplement are as follows:

1. Adding a new specification or test method for a component to ensure better control of quality. Supporting data must be supplied.

2. Changes in artwork text to add or strengthen a contraindication, warning, precaution, adverse reaction, drug abuse dependence, overdosage, or instruction about dosage and administration that is intended to increase the safe use of a product.
3. Deletion of artwork text giving false, misleading, or unsupported indications for use or claims of effectiveness.
4. A new facility or establishment that does not differ materially from the former facility or establishment and has received a satisfactory CGMP (FDA) inspection within the previous 2 years covering the filling or packaging process.

Failure to comply with the supplemental application procedure may result in the FDA withdrawing approval of the NDA. This means that the pharmaceutical company would not be able to continue to market the product in the United States.

4. Annual NDA Reports

The types of change that should be included in annual reports are as follows:

1. Slight modification to noncritical dimensions on a primary component, provided the dimensions do not change the container size (except for solid dosage forms).
2. Change in the manufacturer of a primary component, the polymer, or a component manufacturer (providing the necessary FDA approval of the manufacturer has been confirmed and the DMF reference number has been obtained).

Even if there have been no changes during the year, a report must be sent to the FDA stating that there are no process changes.

5. Drug Master File

The complete file should consist of all of the five types of DMF, although there are only two types of DMF that need to be referred to in relation to packaging components; these are types I and III.

a. DMF Type I—Manufacturing Site and Facilities. This is a general document detailing the manufacturing site, facilities, and personnel.

It is used by the FDA inspector as an aid in finding the pharmaceutical premises (particularly useful for arranging inspections) and in gaining an overview of the factory prior to inspection.

The information that should be included in the document is as follows:

1. Site acreage
2. Address
3. Map showing the location with respect to the nearest city (also an aerial photograph if possible)
4. A diagram showing the site layout, highlighting the actual production areas
5. Details of the major equipment capabilities, applications, and locations (not necessary to quote the make and model of the equipment)
6. Staff organizational structure, highlighting the quality assurance positions. There is no need to name the staff, just the position titles.

b. DMF Type III. This relates specifically to packaging materials. The data that should be included in this document are as follows:

1. Intended use of each packaging material
2. Control for release
3. Names of the component manufacturers
4. Acceptance specifications
5. Data supporting the suitability of the components for their intended use, for example, compatibility data

Sometimes it is difficult to obtain formulation details from a component manufacturer for primary components because the data is highly confidential. In such instances, if a DMF is submitted by the component manufacturer and accepted by the FDA, then this could solve the problem. The component manufacturer then gives permission for the DMF reference number to be quoted in the pharmaceutical company submission documents to the FDA. This arrangement would still require the component manufacturer to give prior notice of any changes to the process to the pharmaceutical company, allowing sufficient time for any work that is required to ensure that product quality is maintained.

III. COMPONENT SPECIFICATIONS

Every detail concerning a component specification must be communicated to and agreed upon with the manufacturer, including packaging, transportation, and labelling requirements. If any of the details are missing, confusion or mistakes may occur.

The main specification requirements are the component drawing, artwork (printed components only), and the quality control testing and standards.

A. Component Drawing

The best method for preparing a component drawing is to use a computer system. This will enable rapid preparation and updating of drawings.

Several rules should be observed when preparing a component drawing:

1. The following details should be stated:

a. Explicit title
b. Specific reference code and version number; the version number must be changed when a drawing is modified
c. Date from which the drawing is to become effective
d. Component specification reference number(s) relating to the drawing
e. Material of construction, grade, and the specific formulation reference number
f. Terminology used, that is, the description of each point of measurement
g. Dimensional limits and units. The dimension limits must be realistic, that is, within the capabilities of the measuring equipment to be used, the component manufacturing process, and the pharmaceutical filling/packaging equipment. Do not quote dimensions at a greater accuracy than the pro-

posed measuring equipment is capable of measuring. For instance, quoting dimensions to the nearest 0.001 in. when the measuring equipment is only accurate to the nearest 0.01 in.

2. A circulation list for each copy of a drawing should be available. Whenever a drawing is updated (i.e., a new version), each copy of the old version should be recalled and destroyed. Failure to do this could result in a component being manufactured using the wrong version of the drawing.
3. Enlarge areas of the component drawing to clarify dimensional details where necessary.
4. Index each dimension with a number on the drawing to prevent mistaking them for dimensional data.
5. The completed drawing must be checked to ensure that all the details are correct. A mistake could cost a lot of time and money.

Observing these rules will ensure that there is complete understanding between the pharmaceutical company and the component supplier concerning the component design, hence minimizing the possibility of errors.

A suitable layout for a molded eye dropper bottle is shown in Figure 1.2, complying with the above-mentioned rules. One point to remember at the design stage, when determining the dimensional limits for a component, is that any overall dimension, such as the thread diameter (9) shown on the neck profile in Figure 1.2, is a combination of the minor thread diameter, wall thickness, and bore diameter. Note that limits are not shown for the wall thickness of the neck, but variations in the other three dimensions can cause an addition of the allowable tolerances in the wall thickness, that is, the minimum, major, and minor thread diameters and maximum bore diameter. Therefore, always make sure that there is still sufficient wall thickness in even the worst possible situation to support the cap and nozzle.

Another example of a component drawing is shown in Figure 1.3. This time, in contrast to a bottle, a carton drawing is used. Note that the position and size of both the bar code and the overprint areas

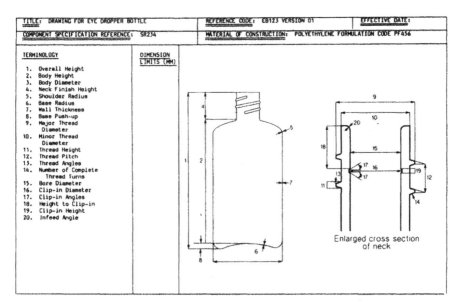

Figure 1.2. Drawing for eye dropper bottle.

are clearly shown. These should also be clearly indicated on the artwork.

B. Artwork

There must be no errors on the completed artwork. The consequences of an undetected mistake could have serious consequences for the customer. The artwork is the prime reference document for the pharmaceutical company quality assurance staff and the supplier. Therefore, the system of preparing the artwork must be carefully thought out to minimize the possibility of errors. Hence, as with drawings, several rules should be adhered to:

1. A specific reference code and version number should be assigned to each piece of artwork. When artwork is modified, the version number must be changed. The reference code and version number must appear on the artwork.

2. A circulation list should be available for each copy of the

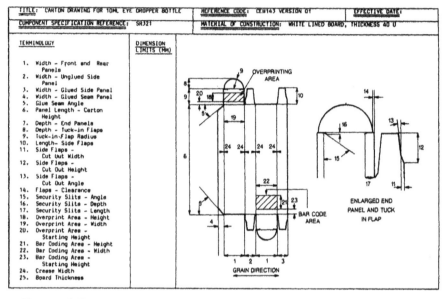

Figure 1.3. Carton drawing for 10 ml eye dropper bottle.

artwork. Whenever the artwork is modified, the old copies should be recalled and destroyed. Copies of artwork will be required by quality assurance and in-process control staff at both the supplier's and pharmaceutical company's premises. If a clearly defined system is not available for the provision of updated artwork, incorrectly printed components may be accepted and used in error.

3. The date at which the artwork is to become effective is required.

4. Suppliers must not be allowed to prepare their own artwork from the master supplied by the pharmaceutical company; this can lead to artwork errors. If artwork is damaged and must be drawn up again, then this must be controlled by the pharmaceutical company, which should ensure that every detail is thoroughly checked. Complicated pieces of artwork with a large amount of print and detail, e.g., leaflets, can take a long time to check and will need checking sev-

eral times to ensure that they are correct. The artwork should be computer generated to ensure rapid preparation and updating.

5. The artwork should be color separated. Providing a supplier with one-piece, several-color artwork means that "blanking out" is required before photographing. Errors can occur during this "blanking out" process, leading to parts of the final text being printed in the wrong color or missing parts of the text. When using color-separated artwork be sure that the registration marks are in exactly the right position, otherwise misalignment of the different colors of text will occur.

6. Stick-on lettering on the artwork should be avoided, as sometimes it falls off if not glued on with a good adhesive.

7. The exact position of the artwork text, color bands, and figures should be shown in relation to the component. This is done by lining in the component on the artwork with a nonreproducable blue ink. The distance from the print starting position should be stated. This is particularly important for bar codes or if an area is to be left blank for overprinting at the pharmaceutical premises. An overprint area that is slightly out of position or too small may cause overprinting problems.

8. The print size and type to be used need careful consideration. Print that is too small may produce legibility problems. Also, print that is close to the edge of a component may result in missing parts of the print due to a slight variation of the print position because of manufacturing limitations.

9. The type and color of ink to be used must be stated, including the lightest and darkest shades allowed. Identical standards should be available in both the supplier and pharmaceutical premises. The ink type to be used will depend on the surface to be printed. Ink that can still be smudged several days after printing is unacceptable. Bar-code readers may only be able to read certain colors, therefore the color to be used requires careful consideration.

10. If any changes are made to the artwork that render the old version unsuitable for use, then all stock of the components printed to the old version of the artwork stored at the supplier's or pharmaceutical company's premises must be destroyed. Be sure to check if any of the components are in transit, collation, or awaiting packaging.

C. Quality Control Testing and Standards

It is necessary to first determine what a batch is for testing purposes. This may be a complete mix for molded components, one shift or a day's work, a whole order for a customer, etc. Whatever a batch is determined to be, it must be traceable back to the raw materials used. If a serious complaint or a product recall occurs, the batch system will help determine the problem source and identify the suspect material on the market.

There are two classes of components:

Primary—in contact with the product, e.g., ampules, vials, plastic bottles, polymer-coated foils

Secondary—not in contact with the product, e.g., cartons, labels, leaflets

Generally the basic testing system is the same for both types of components, although component compatibility and chemical testing are required, in addition, for primary components.

Both primary and secondary components must be tested during manufacture (in-process control) to ensure the best control of quality; this will be detailed in Chapter 3. However, it is still necessary to have clearly defined standards of acceptance/rejection, together with detailed testing procedures.

First, the critical parameters requiring control need defining, and then the standards for each of the parameters.

1. Setting the Standards

There are several distinct areas:

1. Appearance
2. Dimensions
3. Compatibility and customer usability
4. Chemical testing (mainly USP, BS, and DIN testing of primary components, and adhesives on labels for secondary components).

a. Appearance. This can be split into three categories:

(1) Critical. Unacceptable at any level, e.g., rogue printed items in a delivery, incorrect printing of data such as the product name or concentration, insects in the bottles, etc.

(2) Major. Acceptable at a low level; the standard is decided by the pharmaceutical company. It is very easy to ask for perfection, but this is not possible; therefore a reasonable compromise has to be reached. This must take into account the supplier's ability to meet the standard, potential problems during packaging, and whether the defect is acceptable to the customer. Too tight a standard will result in a supplier not supplying because the standard cannot be met or a 100% inspection of each consignment of components received by the pharmaceutical company. This in itself is never a satisfactory solution due to the GMP implications and the fact that inspection is not 100% effective.

Too low a standard could lead to excessive complaints from the market and loss of company image, and hence of orders.

Examples of major appearance defects can be smudged or missing print, making reading the text difficult; flashing on molded components, causing machine handling problems; and other defects, specific either to a certain component or handling process.

(3) Minor. Acceptable at a higher level than the major appearance defects. These will detract from perfection and include marked components, slight color variations, slight smudging, etc.

The best way to set a standard for appearance is to use a statistical sampling and testing system such as Military Standard 105D in the United States (BS6001 in Great Britain). Set a suitable acceptable quality limit (AQL) for the major and minor categories. An AQL must not be set for critical defects, since these are unacceptable at any level, not only during testing, but also when being used by the filling/packaging areas, resulting in immediate rejection of the whole consignment.

A tighter AQL level will usually be set for major defects (e.g., 0.65 AQL) than for minor defects (e.g., 2.5 AQL). Whatever limit is set generally will reflect the highest standard possible. This can be classed as a starting standard, and there is no reason why it cannot be tightened with improvement in the control of component manufacture.

b. Dimensions.   The dimensions of a component can be separated into two types:

(1) Critical.   Requiring close control to ensure that the component functions correctly and can be used satisfactorily by the packaging equipment.

(2) Noncritical.   Necessary to maintain the component shape but not requiring close control (or tight limits) for satisfactory function of the component.

To decide whether a dimension requires critical control or not, the whole pack function and packaging machine requirements require detailed analysis.

A good example to consider is a vial containing an injectable product. The components have to be brought together by the filling machine to give a sterility proof seal, are the vial, rubber plug, and aluminum overseal. Figure 1.4 illustrates the three components and how they form the seal.

The critical dimensions for each of these components are as follows:

1. Vial. Flange depth, flange diameter, bore diameter, vial height, body diameter, wall thickness, base thickness, concentricity, and verticality (and the flange/neck angle with some filling machines)
2. Rubber plug. Flange depth, flange diameter, and plug diameter
3. Aluminum overseal. Internal skirt depth, external diameter, and aluminum thickness (if a plastic, tamper-evident cap is part of the overseal, then the cap diameter and overall height are also critical dimensions)

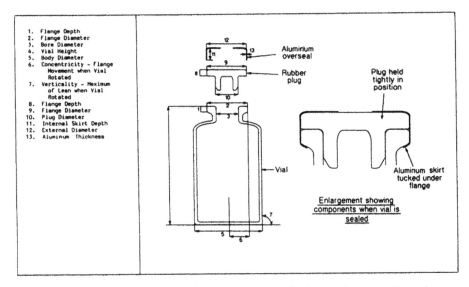

1. Flange Depth
2. Flange Diameter
3. Bore Diameter
4. Vial Height
5. Body Diameter
6. Concentricity - Flange
     Movement when Vial
     Rotated
7. Verticality - Maximum
     of Lean when Vial
     Rotated
8. Flange Depth
9. Flange Diameter
10. Plug Diameter
11. Internal Skirt Depth
12. External Diameter
13. Aluminum Thickness

Aluminium overseal

Rubber plug

Plug held tightly in position

Vial

Aluminum skirt tucked under flange

Enlargement showing components when vial is sealed

Figure 1.4. Cross section of components required to make a sterile seal.

The breakdown of why the vial dimensions tested are critical is as follows:

1. Flange depth. This, together with the plug flange depth, is the total depth that the aluminum overseal has to cover, with sufficient aluminum skirt depth remaining to tuck under the vial flange, yet still hold the plug tightly in place.
2. Flange diameter. Too large a diameter would mean that the aluminum overseal could not fit over the flange. Too small a diameter would mean that the overseal skirt would not tuck under the vial flange.
3. Bore diameter. Too large a bore would mean that the rubber plug would have a loose fit and hence a sterile seal might not be obtained. Too small a bore would mean that the plug could not be inserted.
4. Vial height. The sealing mechanism will not operate satisfactorily if a significant variation in height occurs. A high vial may even be crushed by the filling-machine sealing mechanism.

5. Body diameter. A vial with too large a diameter will not travel down the conveyor track. One with too small a diameter may not align to the sealing mechanism correctly.
6. Wall thickness and base thickness. A wall or base that is too thin may crack or break during washing, sterilizing, or filling. (This also poses a risk to doctors and nurses due to the possibility of breaking during removal of the contents.) A wall or base that is too thick (mainly a problem with molded vials) can also cause problems, particularly if a sterilizing tunnel is used. The vials may be too hot when they exit the tunnel, resulting in rapid cooling, with the possibility of their cracking or being filled with product while still hot. Another risk is the possibility of a vial not being sterile if the safety margins in the sterilizing cycle are very small.
7. Concentricity. This is the amount of flange movement when the vial is rotated about its center and may result in the misalignment of the flange with the sealing mechanism, thus preventing plug insertion and oversealing.
8. Verticality. This is the maximum angle of lean measured from the base when the vial is placed on a horizontal surface and rotated about its center. A leaning vial may result in the misalignment of the flange with the sealing mechanism, thus preventing plug insertion and oversealing. (This fault can also cause labelling problems.)

When the packaging/filling machine brings the three components together, it is necessary for the critical dimensions to be controlled to within very close limits, for example, the vial flange depth, rubber plug flange depth, and internal skirt depth must be such that there is sufficient aluminum skirt for tucking under the vial flange and at the same time for producing tight seals.

(3) Measuring Components. It is not possible to accurately measure components without trained staff and a variety of measuring equipment such as micrometers, callipers, and an optical projector. The variety and types of equipment used are determined by the materials to be measured. There are many types of sophisticated measuring equipment on the market that may be bought for special/rapid mea-

suring of components. Prior to purchasing such equipment, make sure that the equipment is reliable, easily calibrated, and is of known precision and accuracy throughout its measuring range. The more complicated the equipment, the more there is that might go wrong. Equipment such as micrometers are simple to use, reliable, and cheap to buy. A component supplier may not have the financial resources available to buy sophisticated equipment. If measuring differences are to be minimized, it is ideal for the supplier to use the same measuring equipment as the pharmaceutical company.

(a) Measuring Techniques.   Even when measuring something simple with a micrometer, such as the thickness of a sheet of metal, it is possible to measure it incorrectly due either to not using the ratchet or using the ratchet incorrectly. This is because the correct amount of pressure must be exerted by the jaws of the micrometer to avoid errors. The correct method for using a micrometer is to screw the jaws to the work piece very gently by using the ratchet, then, once contact has been made, turn two clicks on the ratchet. A sharp turn, even when using the ratchet, can make the jaws fit too tightly on the work piece and give too low measurement.

Due to the need to measure components of many complex shapes, sizes, and construction materials, accurate measurements can be very difficult, particularly with soft, compressible components. Therefore the equipment to be used and the technique for use must be carefully determined. To do this, it is first important to recognize that a component's shape is not exactly the same as its drawing. An example is shown in Figure 1.5, which shows a close-up of a vial neck and flange profile compared to the drawing.

To measure, say, the flange depth or neck height, an optical projector would be the best choice of equipment. A clearly defined measuring technique must be used. If no technique is given to the operators, six different people would get six different measurements. A reasonable technique for measuring the flange depth is to place the vial upside-down on the projector base plate (see Figure 1.6). Use the base plate as the first reference point A then, using the projector cursors, extend the best line extensions from the bottom edge of the flange to point B. The distance AB will be the flange depth. The technique used for each measurement must be agreed upon with the

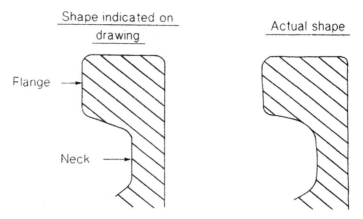

Figure 1.5. Flange profile of a tubular glass vial as shown on the drawing compared to the actual shape.

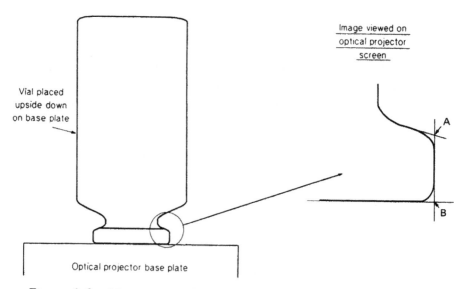

Figure 1.6. Measurement of the flange depth on a glass vial.

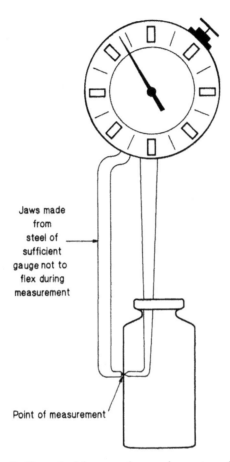

Figure 1.7. Bottle and vial measuring equipment used for side wall measurement.

component supplier, otherwise disagreements will be inevitable, particularly if rejections due to out-of-limit dimensions occur. The measurement of wall thickness could be done with callipers fitted with special jaws to enable measurements to be made (Figures 1.7 and 1.8).

These are inexpensive and effective measuring instruments, but care must be taken when using them. The jaws must be exactly vertical to the wall surface in order to obtain an accurate measurement. An easy way to ensure that this measurement is correct is to angle the jaws

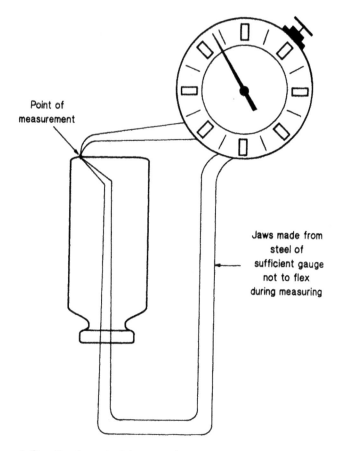

Point of
measurement

Jaws made from
steel of
sufficient gauge
not to flex
during measuring

Figure 1.8.   Bottle and vial measuring equipment used for base thickness
measurements.

slightly to either side of vertical; then the minimum reading will be
correct.

When measuring the base thickness, the minimum thickness
across the base is the required measurement. This may not be possible
with narrow necked vials, particularly when the jaws must be kept
vertical to the glass surface. Several sets of callipers with different jaw
angles will be required to measure the entire vial base. Measurements
at the center of the base can be done with a dial calliper without

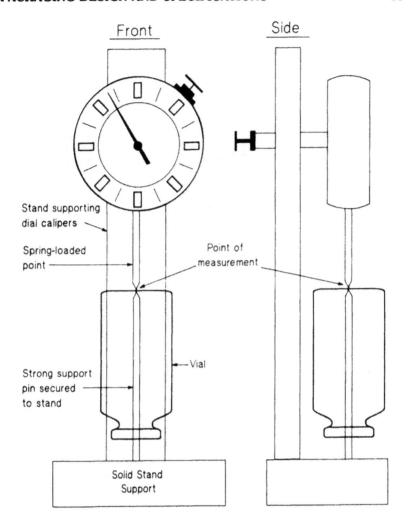

Figure 1.9. Measurement at the center of the base.

specially made jaws (Figure 1.9). With tubular vials the minimum thickness is usually just off the center of the base, with the center being the thickest position. The zero for all these types of callipers must be checked before and after use.

The measurement of soft rubber components such as plugs can be

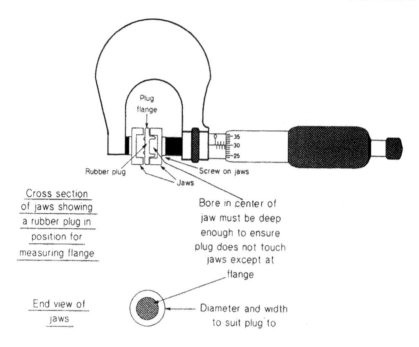

Figure 1.10.   Micrometer fitted with special jaws.

made using a combination of both an optical projector and a micrometer. The diameter can be easily measured using an optical projector, but for flange thickness sectioning may be necessary. This can deform a plug and give false measurements, therefore a micrometer fitted with special jaws should be used (Figure 1.10).

    This minimizes the compressing effects of the micrometer due to the large surface area of contact during measurement. The jaws closely copy the vial flange and aluminum overseal areas of contact and hence give a realistic plug flange depth. It is important to realize that a plug with a varying flange depth will not be detected with this method of measuring. However, provided that control over the making on the molds for molding the plugs is correct (see Chapter 3), varying the flange depth will not be a problem.

    A different set of jaws will be required for each plug size to be measured, taking care that contact is made only with the flange and not with the plug insert or the target ring (if present) on the flange top. The jaws should screw into position (push-on jaws will eventually work loose). The ideal size of a micrometer for this purpose is 25–50 mm (1–2 in.), allowing sufficient room for large jaws and measuring flexibility. The micrometer jaws can then be zeroed on the 25 mm (1 in.) point. This will not be necessary if digital or computerized micrometers are used, since these can be zeroed at any position. It is very important that the jaws be parallel to each other, as otherwise measuring errors will occur. A good local tool-maker should be able to prepare and fit required jaws to a micrometer.

To obtain a specific point measurement on a rubber flange, touch-sensitive measuring equipment should be used. (This will be expensive and will probably be slow to operate.)

It is hoped that the examples of measuring techniques shown have given sufficient understanding of the kind of thought necessary to determine the measuring techniques required. An operator must not just be left to measure a new component before the exact technique has been determined. It is worth remembering that staff who continually measure components will probably have the best ideas for techniques to be used.

(4) Precision and Accuracy. Once a measuring technique is clearly defined, the next aspect to consider is the equipment precision and accuracy. This is basically a question of capability.

First, a set of recently calibrated gauging blocks are required (available from several reputable sources), together with a certificate of calibration. The gauging blocks must cover the full measuring range of the equipment and must be periodically recalibrated at a frequency to be determined by the frequency of use. The gauging block must, wherever possible, be ten times more accurate than the equipment to be calibrated. To check the equipment precision/accuracy, several measurements by several operators (using agreed techniques) will be necessary. The more measurements taken and operators used, the better will be the assessment of the equipment. Let us say that five operators measure each gauge block ten times. This will give sufficient data to assess the equipment. An example is a precision and accuracy assessment of an optical projector (Table 1.1). With this equipment, the

Table 1.1.

OPERATOR REFERENCE	STANDARD BLOCK SIZE (INS.)	INDIVIDUAL MEASUREMENTS (INS.)										MEAN (INS.)	STANDARD DEVIATION
OPERATOR 1	0.2500	0.251	0.250	0.251	0.249	0.251	0.250	0.250	0.252	0.249	0.250	0.250	0.001
	1.5000	1.500	1.500	1.499	1.500	1.501	1.501	1.500	1.498	1.500	1.499	1.500	0.001
	3.0000	3.002	3.000	3.001	2.999	2.999	3.000	3.000	2.999	3.001	3.001	3.000	0.001
	4.5000	4.500	4.500	4.499	4.499	4.498	4.500	4.500	4.501	4.500	4.499	4.500	0.001
	6.0000	6.000	6.001	6.001	5.999	6.000	5.999	6.002	6.001	6.001	6.000	6.000	0.001
OPERATOR 2	0.2500	0.250	0.247	0.249	0.247	0.251	0.249	0.252	0.250	0.248	0.248	0.249	0.002
	1.5000	1.502	1.496	1.500	1.499	1.503	1.502	1.502	1.503	1.499	1.498	1.500	0.002
	3.0000	3.000	3.001	3.003	3.003	3.001	2.999	3.001	3.002	2.997	2.999	3.000	0.002
	4.5000	4.499	4.499	4.500	4.500	4.500	4.498	4.502	4.501	4.499	4.502	4.500	0.002
	6.0000	6.003	6.001	5.999	6.000	6.005	5.999	6.000	5.997	5.999	6.000	6.000	0.002
OPERATOR 3	0.2500	0.249	0.246	0.252	0.251	0.245	0.250	0.253	0.253	0.248	0.247	0.249	0.003
	1.5000	1.502	1.499	1.496	1.503	1.504	1.497	1.497	1.502	1.503	1.500	1.500	0.003
	3.0000	3.001	3.003	2.998	3.001	3.004	2.997	2.997	3.002	3.003	2.999	3.000	0.003
	4.5000	4.498	4.501	4.503	4.504	4.497	4.496	4.502	4.500	4.503	4.498	4.500	0.003
	6.0000	6.003	6.002	5.997	5.999	5.998	5.997	6.003	6.001	5.999	6.002	6.000	0.003
OPERATOR 4	0.2500	0.251	0.252	0.251	0.253	0.252	0.254	0.254	0.253	0.251	0.251	0.252	0.001
	1.5000	1.502	1.505	1.503	1.503	1.501	1.504	1.501	1.500	1.504	1.499	1.502	0.002
	3.0000	3.003	3.003	3.001	3.004	3.002	3.004	3.001	3.002	3.005	3.000	3.002	0.001
	4.5000	4.503	4.503	4.500	4.502	4.505	4.501	4.504	4.500	4.502	4.500	4.502	0.002
	6.0000	6.002	6.003	6.003	6.005	6.001	6.000	6.003	6.004	6.000	6.002	6.002	0.002
OPERATOR 5	0.2500	0.250	0.250	0.251	0.250	0.249	0.252	0.249	0.250	0.250	0.251	0.250	0.001
	1.5000	1.501	1.502	1.499	1.500	1.500	1.499	1.499	1.500	1.500	1.501	1.500	0.001
	3.0000	3.002	3.000	2.999	3.000	2.998	3.001	3.000	3.002	3.002	3.000	3.000	0.001
	4.5000	4.499	4.499	4.499	4.500	4.500	4.501	4.501	4.501	4.501	4.499	4.500	0.001
	6.0000	6.001	6.001	6.000	6.003	6.002	5.998	6.000	6.004	5.999	6.000	6.000	0.001
Overall Mean and Standard Deviation for each block size	0.2500											0.250	0.002
	1.5000											1.500	0.002
	3.0000											3.000	0.002
	4.5000											4.500	0.002
	6.0000											6.000	0.002

operator is an integral part of the measuring system, therefore the precision and accuracy results obtained are a combination of the operator and the optical projector.

Clearly define the rules to be followed by each operator, for example, measure each block in turn and do not measure each block ten times before going to the next one. Also show the operator how to position the crosswires on the edge of the measuring block.

The results in Table 1.1 show that there is a variation between operators, which is to be expected, since no two people will obtain exactly the same results. One interesting aspect of the results is that operator 4 is precise but is constantly reading slightly too high. This indicates that there is a technique problem that needs correcting.

The mean result for each gauging block is a measure of the accuracy, and the standard deviation is a measure of the precision.

The data obtained give an assessment of the equipment under ideal conditions, since the gauge blocks are easy to measure. It is now necessary to consider the equipment precision with a component that is difficult to measure, such as vial flange depth, where special techniques [as described in section (3)(a)] are required to obtain a meaningful reading. It may be argued that the results must be exactly the same as for the standard blocks, provided that each operator uses exactly the same techniques. In reality this is not the case.

One advantage of precision and accuracy assessment is that it can be used for training new operators. An individual using poor techniques would record different precision and accuracy data than the "average" operator.

Once the equipment's capabilities have been determined, it is necessary to ensure that it continues to work correctly by performing calibration checks using the gauging blocks. These checks should be conducted at regular intervals for equipment that is in regular use. The equipment must be zeroed each time it is to be used.

The calibration checks must be recorded, giving details of all problems encountered and actions taken, e.g., repair or servicing by an engineer. An "out-of-calibration" situation must be investigated to ensure that previous measurements performed on the equipment have not resulted in unsatisfactory material being accepted and vice versa.

(5) Measurement Standards. Now that we have identified the

parameters of the component to be measured, the type of measuring equipment, and calibration, it is necessary to decide the number of components to be measured and the standards to be applied.

(a) Molded Components.   The majority of molded components are identifiable by a cavity number. This identifies the mold from which it was produced. Therefore, measuring a few samples from each cavity will be sufficient. The frequency of measuring samples will depend on the rate of wear of the mold. A mold that lasts 10 years is going to show little sign of wear from month to month, whereas a mold for, say, a polyvinyl chloride (PVC) component, may wear out in a few weeks (unless specially coated).

It is very important that with refurbished molds (due to damage etc.) or new molds, samples from each cavity **must** be checked before putting the mold into production use. Also, a different formulation of polymer or a different polymer used for a component in the same molds (tooling) may give different measurements due to different rates of shrinkage. Therefore, the measurements must also be checked with such changes.

It is worth noting that many suppliers may produce their molds to allow maximum usable life with respect to wear. This will mean that molds will be produced to meet the minimum component dimensional standards.

The dimensional standards to be applied for molded components should be "all dimensions to comply with the drawing limits."

This standard may appear to be very strict and possibly, based on one's own past experience, impossible to achieve, but if the aforementioned actions are taken, this standard is possible to achieve. However, some relaxations may be allowed for noncritical dimensions, e.g., the shoulder profile on a bottle. It must be remembered that for every component with a critical dimension outside the agreed limits (provided they have been set correctly), an equipment stoppage may occur or it may result in a defective pack on the market. Therefore, tight standards are necessary.

(b) Nonmolded Components.   Glass ampules, tubular vials, and collapsible aluminum or laminated tubes are all nonmolded components. A statistical sampling system using Military Standard 105D (USA) should be used with these components. This should only be necessary as a check to ensure that the in-process control (IPC) system at the suppliers' premises is working satisfactorily.

The sample size taken for dimensional checks should be related to the IPC system used by the supplier. A zero AQL standard should be used for critical dimensions.

(6) Computerization of Measuring Equipment. With all the technology currently available, it is very easy and reasonably inexpensive to interface measuring equipment to a computer and printer, e.g., callipers, micrometers, and optical measuring equipment. There are several advantages to computerization:

1. Prevention of operator transcription errors
2. Rapid recording of all results
3. Audible indication when results are outside limits (if required)
4. Instant computer statistical calculation and printout of results, and a statement of whether the consignment is acceptable
5. Computer can be programmed to print out in a standard work-book format that incorporates the results
6. Securing against shortcuts that some operators may be tempted to take; for example, it can be programmed to include a zero check and calibration
7. Inclusion of measuring instructions; for example, on entering the component code and batch number, the computer might automatically inform the operator of the dimensions that need to be measured and how to measure them (this would probably only be required in the training of new staff)

c. Compatibility and Customer Usability. This involves checking that each component forming a pack fits together and functions correctly.

Consider an eye dropper pack as an example. The nozzle must have a good interference fit into the bottle and allow one-drop-at-a-time delivery through the hole in the nozzle when inverted, but must not leak from the fitted position. The cap must screw into position, and leakage must not occur when the bottle is squeezed in the inverted position, i.e., a sterile seal is maintained. The bottle can be filled with water for this trial.

The cap must be capable of being removed easily by the customer. This should be checked by screwing the cap into position at the torque operated on the filling machine, then removing the cap by hand. Sometimes backing off of the torque may occur; this should be investigated at the design stage. If this is a problem, then the time taken for the backing off to cease is the period the caps should be left in place before removal. Trials should be performed with several combinations of components produced on different mold cavities if realistic results are to be obtained. The more cavitites there are for each component, the more samples are required.

When a pack has been designed to be child resistant or tamper evident, these parameters should be checked to ensure that they function correctly.

d. Chemical Testing. The majority of chemical testing is required on primary components. The main reasons for this are to ensure that the correct construction materials have been used at the required standard for manufacturing the components. Product degradation or contamination may occur if wrong construction material or unacceptable contaminants are present.

The type of testing required depends on the type of component used.

(1) Glass Vials and Ampules. The USPXXII requirements for glass containers are chemical resistance and light transmission (only necessary for light-resistant containers). The requirements vary from country to country, but basically testing determines whether the correct type of glass has been used for manufacture and its suitability for use with pharmaceutical products.

(2) Plastic Primary Components. The testing is more extensive with plastic components, requiring both biological and physicochemical tests. This is because plastic components contain other substances, such as plasticizers, stabilizers, antioxidants, pigments, lubricants, and possibly residues from polymerization. Therefore, for components that are in direct contact with the product (and in particular liquids), this

testing is required to help ensure that the product is not affected during its life.

Each batch of components will require testing to ensure compliance with the chemical standards.

D. Component Specifications Layout

There is a tremendous amount of information to include in the specifications. All changes must be agreed upon with and communicated to the supplier(s). There may be regular changes to artwork, testing methods, drawings, and additions/deletions of component codes covered by the specification. Therefore, rather than having to reissue the whole specification when any of the above changes are necessary, it is suggested that component specification be split into four parts, as follows:

1. The general specification, which states the appearance standards, labelling requirements, packaging and transportation details, and laboratory testing standards and techniques, as well as any other items that are not likely to change very often. This will cover the basic data relating to possibly several component codes. This general specification should have a reference number.
2. The artwork should be a separate document, with a reference number included on the artwork.
3. The drawing for a component should also be a separate document, with a drawing reference number.
4. The testing methods for both chemical and dimensional checks will be general documents, relevant possibly to a whole class of components e.g., cartons, vials, labels. It is still not wise to incorporate these in point 1, because they may require minor modification, e.g., a change in the equipment for performing a task. These will possibly vary slightly for use in the supplier's premises compared to a pharmaceutical factory, due to variations in the type of equipment used, e.g., different types of optical projectors. However, the measuring and laboratory techniques must be identical and should be included in the core part of the specification (point 1).

COMPONENT DESCRIPTION   QUALTAB CARTON

COMPONENT CODE   C101 VERSION 01

GENERAL SPECIFICATION
REFERENCE   GS012 VERSION 01

ARTWORK REFERENCE   A201 VERSION 02

DRAWING REFERENCE   DR213 VERSION 01

TESTING METHOD   TM034 VERSION 01

EFFECTIVE DATE   (Date the Specification is
                                                       to become effective)

AUTHORISED SIGNATURE

This is an important official document and
requires authorization from a senior member
of the
Pharmaceutical Company's staff
(not necessarily a Quality Assurance Person)

Figure 1.11.   Specification tie-up document.

The testing methods should have a testing method reference number.

The splitting of the specification into four separate units makes updating much easier. A specification tie-up document, as illustrated in Figure 1.11, has to be issued both as part of the initial specification and when modifications are made. A version number must be included with the general specification, artwork, drawings, and testing methods, because sometimes changes will occur that do not require the issue of a new component code, e.g., minor dimensional changes, different method of packaging components, and the correction of minor mistakes in the artwork, such as the repositioning of wording.

Standardization with a basic component design will make this layout and system of issuing a specification all the more useful.

2

Supplier Quality Auditing

I. INTRODUCTION

A quality audit determines whether a new supplier is basically suitable for producing components to the standard required or whether an existing supplier is continuing to produce at the standard required. Mistakes at this stage can have serious implications both from a quality and financial viewpoint.

It is therefore essential that a quality audit is performed in the correct manner by a well-trained quality auditor.

II. QUALITY AUDITOR

Considering the complexity of quality auditing, the different types of industries involved and the limited time period for performing the audit, the quality auditor must have many specific personal qualities as well as special training.

A. Personal Qualities

1. The auditor must be a realistic, practical person capable of quickly understanding the process details and the practical problems encountered by the supplier. This will ensure that unrealistic demands are not made upon the supplier by the auditor.
2. The auditor must be capable of communicating with staff at all levels, from production operator to senior manager. This is necessary for the auditor to fully assess the process and staff.
3. The auditor must be very observant and be prepared to ask questions about and ask to see areas and equipment bypassed during the tour of the factory.

B. Training Requirements

1. Extensive experience with the use of the components in pharmaceutical premises, in particular potential quality problems and standards operated, is necessary.
2. Full awareness of the GMP requirements for component manufacture and usage. Also be aware of the legal regulations and the particular country requirements in which the finished product will be sold.
3. Experience of the component manufacturing process prior to an audit is required. This will ensure that none of the manufacturing stages are omitted during the audit.
4. Visits to premises manufacturing similar components are required to obtain experience with the general GMP standard for the industry. Not having this experience could lead to unrealistic standards being requested of the manufacturer. To try to bring all suppliers up to pharmaceutical manufacturing standards immediately would be very difficult and probably financially impossible from the supplier's point of view. A

gradual improvement of premises, facilities, and staff GMP training can be more reasonably expected.

A summary of the quality auditor requirements is shown in Figure 2.1.

III. AUDITING

All suppliers should be quality audited, but the areas to concentrate on first are those supplying primary and printed materials.

The quality auditor must prepare in advance of the visit. First determine the main business of the company. Find out whether they normally supply the pharmaceutical industry or a less critical customer. A company that has never supplied a customer with strict GMP standards may require reeducating from the management down to the operator level. This is a mammoth task and will take a lot of special attention by the pharmaceutical company. Determine if nonstandard production is being requested, i.e., very small order quantities when they usually produce very large quantities (dedicated lines). This would highlight two potential problems:

1. Line changeover procedures will require special attention, particularly with respect to cleandown and reconciliation to prevent rogues.
2. A forecast production system may be operated, i.e., 6 months or a year's predicted offtake would be produced at a time to minimize costs; for example, production of molded bottles using a special glass. This would necessitate the supplier storing stocks for a considerable length of time. Hence the packaging and warehouse would need special attention because of the possibility of vermin contamination or deterioration during prolonged storage.

Confusion can easily occur if a set auditing sequence is not followed, leading to important areas being missed.

It is always best to start at the beginning, i.e., at the raw material storage area, and follow through the process in the manufacturing sequence to the final dispatch to the customer, covering all aspects of

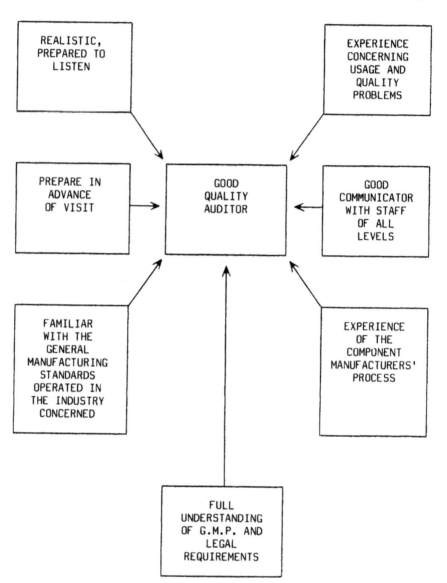

Figure 2.1. Quality auditor requirements.

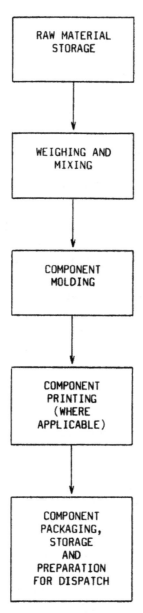

Figure 2.2.   Auditing sequence—Rubber and polymer components.

GMP at each stage. Figure 2.2 shows the auditing sequence for rubber and polymer components.

A. GMP Requirements for Raw Material Storage

There can be a wide variety of materials, considering the many different types of components that can be used by pharmaceutical companies, but there is only one main aim—to ensure that the raw materials are stored correctly, do not deteriorate, and are not contaminated prior to use. Special storage conditions may be required for some materials.

The main GMP considerations are

1. A building of good, solid construction and design to minimize vermin infestation, i.e., birds, insects, mice, etc. The large access doorway to the storage area is usually the main entry point for such infestation, therefore, this should be kept closed as much as possible. In addition, regular inspections and control measures should be carried out, using a pest control company that applies approved pesticides.
2. The building should preferably have no windows, as sunlight can deteriorate, discolor, or fade materials.
3. It should have a sealed concrete floor or similar material that minimizes dust generation from fork-lift trucks, etc.
4. Open drains must not be present in the warehouse. This would be a potential bacteriological risk that might contaminate the raw materials.
5. Extremes of heat, cold, and dampness should be avoided. Air conditioning may be necessary to prevent the deterioration of some materials.
6. Adequate segregation of different materials to prevent possible mix-ups, damage, or contamination (e.g., material with a strong odor) is necessary. Liquids should be stored at the ground level, with a catchment area in case of spillage. No items should be stored in direct contact with the ground. High rack storage should be used to make the most efficient use of space available and to prevent damage of materials from placing one full pallet on top of another. An over-

crowded storage area can create the above-mentioned problems and can make access difficult. Stock rotation may not always be possible as a result. The area must be kept clean and tidy.

7. There must be an organized storage and stock control system to ensure correct stock rotation, i.e., the oldest stock must be used first. Each raw material must be reassessed if not used within a certain time as agreed to by quality assurance.

8. Status labelling and quarantine areas should be set aside for storage of materials awaiting testing and rejection. This is a sure way of ensuring that such materials are not used by mistake. Pass labels must state the reassessment date. If a computer control system is operated instead of status labelling, then this must be proven to be effective in the control of under-test, passed, and rejected materials.

The GMP requirements for raw material storage are summarized in Figure 2.3.

All raw materials to be used for pharmaceutical primary components should be received with a certificate of analysis/conformance or be sampled and tested on receipt. Sampling should be done in a clean area or sampling booth.

B. GMP Requirements for the Formulating Area

1. A dedicated clean area must be available for weighing and mixing materials.

2. Authorized formulation procedures must be available and followed.

3. Only **one** formulation is to be weighed and mixed at a time in any one area to prevent mix-ups and cross-contamination.

4. The container into which each material is weighed must be clearly labelled. Each weighing operation and addition to a batch should be checked by another operator or supervisor. These operations should be recorded on a batch sheet.

5. Ideally, computerized formulating instructions and check-weighing systems should minimize potential human errors.

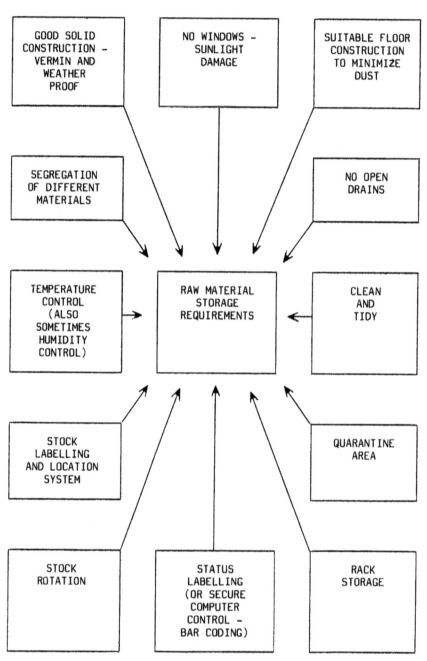

Figure 2.3.   GMP requirements for raw material storage.

These could also be connected directly to the ordering system in order to minimize production delays.

6. Weighing equipment must be calibrated regularly with standard weights and should have a calibration certificate that is traceable to the National Bureau of Standards (or the relevant country equivalent for overseas suppliers). Ensure that the weighing equipment is calibrated over the whole of its weighing range. A record must be kept of these checks.

7. Each operator and supervisor must be fully trained with a full appreciation of the GMP aspects of the job.

The GMP requirements for the formulating of rubber and polymer components are summarized in Figure 2.4.

C. GMP Requirements for the Production of Components

Whether molded bottles or plugs, or printed cartons or bottles are involved, there are several rules that need to be strictly adhered to in order to ensure that good quality is maintained.

1. Each machine must be separated by a barrier or at least sufficient space to ensure that neither materials nor staff overlaps can occur. Ideally there should be a separate room for each machine. This is to prevent cross-contamination between machines, i.e., rogues. Primary components should be produced in a clean area.

2. Staff allocated to one machine must not be allowed to wander freely between machines. Again, this is to prevent rogues.

3. Prior to the start of production, a check must be made to ensure that everything is correct. The machine and area must be thoroughly cleaned and completely free from materials used or produced on the machine previously, e.g., the correct molds should be fitted to the molding machine with the correct batch of polymer mix, the correct print text and colors with the printing machines, etc.

    The above actions are required to prevent rogues, cross-

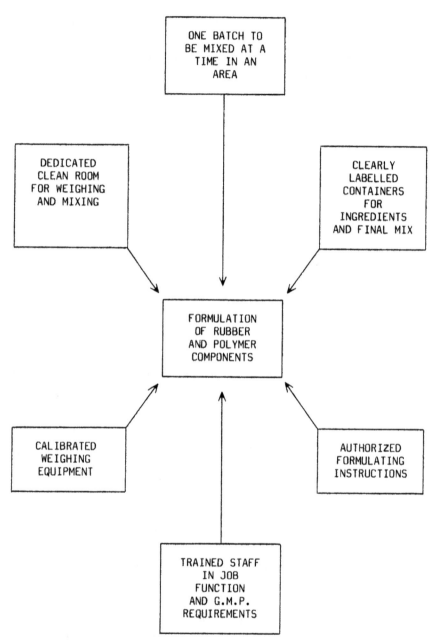

Figure 2.4. GMP requirements for the formulation of rubber and polymer components.

contamination, and the use of incorrect materials during production.

4. Machine operators, supervisors, quality assurance staff, and
   engineers must be fully trained by experienced personnel,
   and this training should include the GMP aspects of the job.
   Records of all training must be kept.

   Putting a new member of staff straight onto an operating
   line without training could have disastrous results on output
   and quality.

5. An in-process control (IPC) system should be operated on
   each production machine or line. This is the most effective
   way of controlling quality. Basically this involves regular
   monitoring of quality throughout the manufacturing operation. All IPC checks must be recorded. The first samples
   produced must be checked.

6. The output from each machine must be placed into clearly
   labelled containers. The labels for these containers must be
   prepared (ideally, computer generated as required) in a secure area and be accurately reconciled. The label should
   state material name, reference code, batch number, quantity,
   date produced, and operator's name.

   With primary components, special packaging may be required to minimize contamination during transportation,
   i.e., nonfiber shedding materials, double bagging, packaging under clean conditions. Such precautions can minimize
   cleaning problems for the pharmaceutical company.

7. Each batch produced from a machine must be quarantined
   until considered suitable for passing by quality assurance.
   With a good IPC system, this will probably be when the
   production run has finished, and hence it can be passed
   immediately. Otherwise the batch must be placed in a quarantine area or at least be labelled *quarantined.*

GMP requirements for the production of components are summarized
in Figure 2.5.

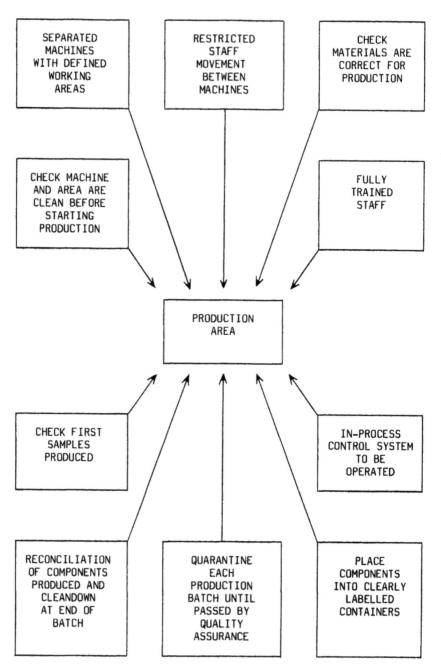

Figure 2.5. GMP requirements for the production of components.

D. Finished Product Storage Area

This must be of the same standard as the raw material storage. Each order must be kept separate, and preferably each batch in an order should be kept separate, e.g., on a separate pallet. Rack storage should be used. Loaded pallets should not be stacked on top of one another unless the component packaging has been **designed** to take the weight.

E. Completion of the Audit

When the audit has been completed, the auditor must prepare a report detailing all the GMP practices, both good and bad, observed during the audit. A copy must be sent to the supplier for comment. The follow up to this will be regular communication and cooperation with the supplier to resolve any GMP problems observed. It is the decision of the quality auditor as to whether a supplier is acceptable or if GMP improvements are necessary before acceptance. An official list of all approved suppliers must be kept by the pharmaceutical company.

Each approved component supplier must be audited at regular intervals to ensure that the quality standards have been maintained. This should be every 1 to 2 years, or whenever a serious problem is encountered.

IV. PROBLEMS ENCOUNTERED BY AUDITORS

Like many things in this world, quality auditing does not always go as planned. The following are some of the situations that may be encountered:

1. Pharmaceutical company requirements are sometimes very small compared to a supplier's other customers, therefore, suppliers may not be prepared to improve their standards to suit the pharmaceutical industry (at any cost). This usually happens when there are no other suppliers of a particular item. Until an alternative supplier becomes available, the quality will have to be built into the product by the pharmaceutical company, e.g., extra washing, 100% inspection.

This situation is far from satisfactory, but, provided it is dealt with correctly, will result in a satisfactory component. With 100% inspection, for example, the staff must be well trained and have a dedicated secure area to work in. The inspection efficiency must be known, and the inspected materials must be statistically sampled and checked by quality assurance.

2. Suppliers cannot afford to bring their manufacturing premises/processes to the necessary standard. This is a similar situation as above and the same actions apply, but alternatively the pharmaceutical company could provide financial aid or equipment to the supplier.

   The pharmaceutical company could either buy a company or set up to produce components themselves if situations mentioned in points 1 or 2 arise. This would enable the components to be manufactured to the GMP standards required.

3. Suppliers not following the manufacturing process through in logical order during the audit. This can cause confusion for the auditor, who will possibly miss an important area. A likely situation for this to occur is when the next stage of the process is at the other side of a factory and a later stage is nearer. In this situation it is best to insist on following the process in logical order, as this gives an indication as to how well the factory is organized.

4. Suppliers trying to keep auditors from problem areas, by spending too much time in good areas, hoping there is insufficient time to see the low standard areas. A strict timetable must be adhered to, particularly if only a limited amount of time is available.

5. Spending too much time around the conference table or at lunch, leaving less time to spend on the audit. Again this comes down to setting a strict timetable at the start of the audit.

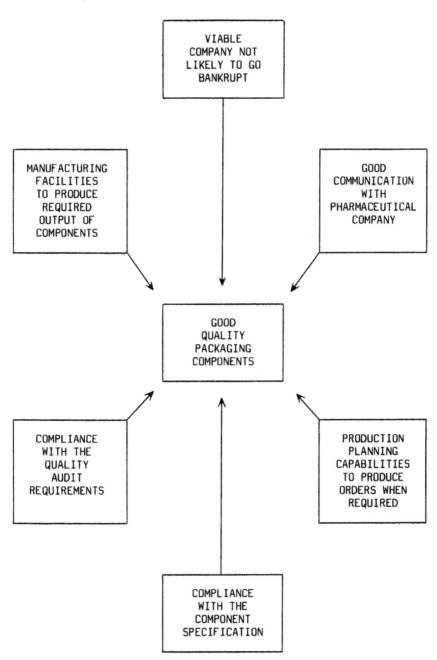

Figure 2.6. Supplier requirements.

V. GENERAL COMMENTS
   CONCERNING SUPPLIERS

It is important to have a good working relationship with a supplier, and a carefully thought-out purchasing policy will help to obtain this aim. Whether a system of single, dual, or multiple sourcing of components is followed can have a significant effect on the quality obtained. It is my opinion that single sourcing will give a greater sense of commitment to the supplier and hence incentive for full cooperation with the pharmaceutical company. However, the supplier must be capable of the present and future production requirements. If a supplier tries to produce at a rate that is higher than the equipment is capable of handling, this will inevitably lead to quality problems. It is also necessary to ensure that a supplier is commercially viable, as a company suddenly going out of business may cause a serious component shortage problem. This is an instance in which having more than one supplier is an advantage, but, considering the extra work involved in having to audit more than one supplier of a component and to set up the communication system, etc., this increases the risk of quality problems. If a dual sourcing system was operated and a total of 40 different component-manufacturing facilities were required, this would mean identifying 80 suppliers. All would have to be audited, possible quality problems would have to be resolved, and a communication network would have to be set up. This would require extra resources with respect to quality auditors. It is a lot easier to control the quality from 40 rather than from 80 suppliers.

Figure 2.6 shows the basic supplier requirements for supplying the required quantity of high-quality components by the agreed delivery date.

3

Quality Control at the Supplier's Premises

I. INTRODUCTION

Now that we have both the specification and a supplier identified, continual communication and cooperation is required between the pharmaceutical company and the supplier. Therefore, a contact person for each company must be identified, i.e., a senior quality assurance person. This must be the only route of communication concerning quality matters. Sales staff/buyers are usually not technically competent enough to discuss quality aspects and could cause misunderstanding or confusion. Ideally the contact person for the pharmaceutical company should be the quality auditor, who will be more familiar with the supplier's processes than anyone else.

The aspects to be considered are the GMP deficiencies identified during the audit, the component specification, and the mechanism for the two-way communication of quality problems.

II. GMP DEFICIENCIES

The main areas likely to be deficient during the quality audit are

1. Staff training
2. In-process control during component manufacture
3. Production operation security
4. Good housekeeping
5. Tooling control

The approach required to help resolve these problems will be covered in this chapter. It can only be adequately achieved by having the pharmaceutical company representative spending a great deal of time on the supplier's premises. This is the best way of achieving exactly what is required and hence of achieving consistency between suppliers.

A. Staff Training

Staff that are inadequately trained **will** make mistakes that may lead to quality problems. Putting a new member of staff alongside another operator or technician is not sufficient. An experienced staff trainer must be available, someone who is a good communicator and can monitor the training process. Training in some instances can take many months. One important point relevant to all members of staff is that everyone must be quality conscious. Quality must not be sacrificed to increase or maintain output. All staff associated with the production process must be trained, including the service engineers and quality assurance staff.

It is easy to say what the training requirements are, but to ensure training is completed adequately can be difficult. Sitting staff in a lecture theater or classroom and lecturing on the training requirements is not satisfactory. It is likely that very little will be learned and what has been learned may not work in practice in the production area. A combination of both interactive classroom sessions and on-the-job training is necessary. The only proof of being trained is via monitoring of the on-the-job training by an experienced training officer.

Close monitoring is also necessary to ensure that no mistakes are made. Ideally, a production line/machine specifically used for training would be perfect, but, due to lack of space or resources, this may not be possible.

When new equipment is installed, the training of all staff must be organized and completed before production is started.

Training records must be kept for all members of staff, indicating the date trained, the time taken to train, the signature of who performed the training, and the job function in which the person was trained. These records are required by the relevant immediate supervisor to ensure that the right staff (i.e., trained) are chosen for a particular job, whether refresher training is necessary, or if training in a particular area has been missed.

Production, engineering, and quality assurance staff have different training requirements. These have been listed in Tables 3.1 to 3.4.

Standard operating procedures should be prepared for all quality assurance, production, and engineering activities. These will serve two main purposes, ensuring that job functions are clearly defined and that they can be referenced by the relevant staff if they are unclear on how to perform a task. The procedures are considered to be training documents. Even when someone is fully trained, certain job functions may not be performed regularly enough to be remembered, e.g., analysis of materials, maintenance, or repair of equipment. This is particularly true with reference to analysis, where procedures have probably always been available.

B. In-Process Control During Component Manufacture

1. Definition

Controlling the quality at the time of producing the material is necessary, rather than assessing the quality of a batch of material after it has been produced.

Table 3.1.   Training of production/machine operators.

TRAINING REQUIREMENTS	DETAIL	POSSIBLE CONSEQUENCES OF NON COMPLIANCE
Machine Operation	The operator must know how to set up and operate machine correctly (this includes safety training) Also what action to take when a problem occurs	Defective components may be produced. Almost certainly the highest possible standard will not be obtained. Probably high reject rate and low output efficiency. Delay in supplying Pharmaceutical Company. Batch may be rejected.
Quality Standards	Fully aware of all likely defects and the acceptance standards. Ensure machine fault is corrected immediately. Defective material must be removed from machine. The supervisor must be informed when defects are found.	Defective components sent to Pharmaceutical Company, complaint or possible rejection
Security/Good Housekeeping	Check the correct materials are received and the correct quantities. Reconcile output against input quantities. If printed components check print is correct against the artwork.	Possibility of wrong material used - could affect Pharmaceutical Company's final product quality or reduce line efficiencies. Incorrect print could be fatal to a patient - batch recall by Pharmaceutical Company.
Security/Good Housekeeping Cont/	Machine and area must be thoroughly cleaned when a production run is completed <u>before</u> the next production run is started. The operator must know where to search the machine and area for rogues	Possibility of contaminated components, or rogues, in next production run. Product degradation or rogues on market if not detected - possible product recall - possible death of patient.
	Operators must not be allowed to move freely from one machine to another producing different components.	Possible transfer of components between lines - i.e. rogues.
	Protective clothing must be worn correctly at all times.	Possible particulate contamination of components.
Rejects	All on line rejects to be immediately placed in a clearly labelled reject container. Components dropped on the floor must be considered as rejects	Possibility of faulty components getting among good components. Complaint or rejection by Pharmaceutical Company

Table 3.2. Training of quality assurance staff.

TRAINING REQUIREMENTS	DETAIL	POSSIBLE CONSEQUENCES OF NON COMPLIANCE
Quality Standards	Quality Assurance Staff involved in either In-process Control or batch/raw material testing must be fully aware of the standards to be applied and actions to be taken if/when non-compliance occurs.	
Testing	Quality Assurance staff are no different to either Production or Engineering staff concerning training on how to perform a particular job function. Whether analyzing a raw material, measuring a component or sampling a batch, the method and technique must be known.	Batch may be passed as suitable for sale when it is unsuitable. Likely to be rejected by the Pharmaceutical Company.
Security/Good Housekeeping	All areas of the production process that may cause a security problem must be known and monitored. Quality Assurance must "police" Production and Engineering areas to ensure compliance with Factory G.M.P. standards.	

Table 3.3. Training of engineering staff.

TRAINING REQUIREMENTS	DETAIL	POSSIBLE CONSEQUENCES OF NON COMPLIANCE
Changeover and Machine Set Up	The mechanical Engineer must know how to assemble and set up each machine. The parts must be carefully examined to ensure they are not damaged.	Poor quality materials produced High level of wastage or batch rejection. Possible machine breakdown, low output. Delay in supplying Pharmaceutical Company.
Maintenance, Calibration and Repair	This can involve mechanical, electrical and electronics engineers. All must be fully trained in each of these job functions.	As above.
Quality Standards	All engineers involved in the set-up or repair of production equipment must be fully aware of all likely defects and the acceptance standards.	Defective components may be produced. Possible conflict between production staff and engineers. Poor quality components may be sent to Pharmaceutical Company.
Security	Any packaging items used in setting up or checking equipment after repair must be removed from the machine, and rejected/rogue items found within the machine must be handed to the machine operator/supervisor for reconciliation and rejection. If possible, modifications should be made to prevent rogues getting inside the machine. Tool boxes must be kept clean and tidy.	Incorrectly labelled product on market could harm a patient. Recall may be required.

Table 3.3 Continued.

TRAINING REQUIREMENTS	DETAIL	POSSIBLE CONSEQUENCES OF NON COMPLIANCE
Good Housekeeping	Equipment must be cleaned when an engineer has finished working.	Marked material may be produced resulting in rejection
	Electricians or plumbers working over equipment producing packaging components must ensure machine is protected from particulates etc., e.g. dead flies and dust from light fittings. Covering with a dust sheet is necessary.	Contamination of materials. Possible contamination of product when components used by Pharmaceutical Company. Such instances usually contaminate a small number of items and therefore are likely to remain undetected until a customer complaint is received. A dead fly found in a pharmaceutical product can considerably damage a Company's image. Metal particles in eye preparations may damage a patient's eyes.

Table 3.4. Training of production supervisors.

TRAINING REQUIREMENTS	DETAIL	POSSIBLE CONSEQUENCES OF NON COMPLIANCE
Quality Standards	Fully aware of all likely defects and the acceptance standards. Ensure staff report any defects found. Ensure corrective action is taken immediately. Investigate the extent of the problem. Ensure all suspect material is quarantined. In-process Controller and Q.A. to be informed.	Defective components may be sent to Pharmaceutical Company, complaint or possible rejection.
Security/Good Housekeeping	Ensure correct materials are received and the correct quantities. Ensure the output is reconciled against input. Investigate any discrepancies found. Ensure what is produced is exactly what has been requested, e.g. correct bottle size, print etc.	Possibility of wrong material used - could affect Pharmaceutical Company's final product quality or reduce line efficiencies. Incorrect print could be fatal to patient - batch recall by Pharmaceutical Company.
	Ensure production machine(s) and area are thoroughly clean when production run is completed before the next production run is started	Possibility of contaminated components, or rogues, in next production run product degradation or rogue on market if not detected - possible product recall - possible death of patient.
	Staff must not be allowed to move from one machine to another producing different components.	Possible transfer of components between lines - i.e. rogues.
	Ensure staff wear the relevant protective clothing correctly at all times. The required level of personal hygiene must be maintained among the staff.	Possible particulate contamination of components.

Table 3.4.   Continued.

TRAINING REQUIREMENTS	DETAIL	POSSIBLE CONSEQUENCES OF NON COMPLIANCE
Rejects	Ensure all on-line rejects are placed in a clearly labelled reject container. Components dropped on the floor must be considered as rejects.	Possibility of faulty components getting among good components. Complaint or rejection by Pharmaceutical Company.
Supervisor/Operator Communication	It is important that there is a continual communication between staff and supervisors. Staff must inform the supervisor immediately of any problems or mistakes. Covering up mistakes may lead to serious problems to the customer if not detected. The supervisor must ensure that his/her staff are fully aware that a genuine mistake will not lead to discipline (this is the main reason why people cover up mistakes) unless a deliberate act of disobeying the rules. A mistake quite often means there is a problem with the system that needs investigating and correcting. The supervisor must gain the respect of his/her staff in order to set up this line of communication.	Lack of communication can mean the supervisor is not fully aware of what is happening in his/her area. This can result in poor decisions being made and a lack of awareness of staff needs. Also if mistakes are not communicated to the supervisor, poor quality materials/rogues may be sent to the Pharmaceutical Company. It is unlikely that improvements in the production operation will occur without this two way communication.
Attention to Machinery	The supervisor must ensure that 100% attention is given to running machinery. This will help reduce quality problems and machine breakdowns. For example a jam up on a fast running production line will possibly cause serious machine damage unless observed and acted upon quickly.	Poor quality components, excessive machine down time and a high level of rejects.
Training	The supervisor must ensure all his/her staff are fully trained for the job they are doing.	Poor quality components, machine breakdown, high level of rejects and low output.

2. Advantages

The advantages of in-process control (IPC) compared with batch testing are

1. Only a small part of the output produced (compared to possibly a day or more's production with batch testing) is at risk if the quality is unsatisfactory.
2. Less poor-quality material is produced because the problems are detected quickly and hence corrected; this gives a higher output and is therefore a cost saving.
3. Material is suitable for dispatch to the customer as soon as the order is completed (no waiting for testing to be completed, with the added possibility of rejection). Hence the order is received by the customer more quickly.
4. There will be sufficient assurance of batch quality that it will not be necessary for the pharmaceutical company to test the components. (In-process testing protocols will be received with each batch.) This will have further advantages in that there will be no time delay in testing the components when received. Hence the storage time will be minimized, thus making economic use of the space available.

The advantages of in-process control during component manufacture compared to finished batch testing are summarized in Figure 3.1.

3. Equipment Capability

Before in-process control can be set up, the production equipment capability must be determined. If this is not done correctly, the IPC system will not ensure effective control of product quality.

Determination of the equipment capability involves a detailed analysis of all aspects of the process to determine all the parameters that can give rise to poor product quality. There should be background data to help assess these parameters, such as batch rejects, and other quality assurance analysis data. If insufficient data is available, an extensive sampling system is required at each stage of the production process, examining all samples carefully for all quality parameters,

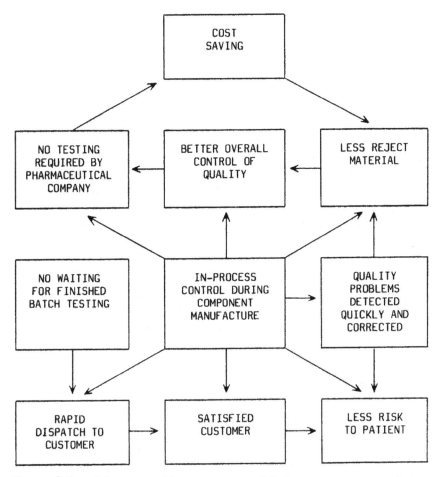

Figure 3.1.   Advantages of in-process control during component manufacture compared to finished batch testing.

e.g., appearance, dimensions. This will have to cover at least several days' production (possibly a few weeks) to take into account variables such as different staff and raw materials.

What will come out of the exercise is a list of all the quality problems that are usually encountered, the part of the production process producing the defects, and the quantity of defects produced at the operating speed(s) used. This is the basic data required to set up the in-process control system. A further advantage is that particularly troublesome areas may be identified that may be possible to correct, for example, by simple equipment modifications.

4. In-Process Sampling and Testing

The first step is to decide the stages of the production process where sampling for quality checks is required, the number of samples, the frequency of sampling, and the quality parameters to be checked.

Appearance is usually the hardest quality parameter to control and therefore requires more monitoring. The number of samples taken, and possibly the frequency of sampling, will depend on the machine speed; usually the faster the machine, the poorer the quality. A statistical sampling method must be used, i.e., Military Standard 105D in the United States (British BS6001). The sample taken must be random over the output produced since the previous check. The sampling plan used and standards operated must ensure compliance with the component specification agreed with the pharmaceutical company.

For example, suppose the AQL for major defectives is 0.65 using General Inspection Level II and an output per 10-hour shift of 10,000 items. If the quantity produced during the shift is considered to be a complete batch, the sample size would be 200 if end-of-batch testing was to be done and not in-process testing.

If it is agreed to in-process sample once per hour, an AQL of 0.65 using Special Inspection Level S4 for an output of 1000 items per hour would require a sample size of 20. This would give a greater assurance of control than batch testing.

One very important point with in-process testing is that if the components are considered to be rejectable, all output since the last in-process check must be isolated. This means that it must be possible to identify all material produced since the last check. The easiest way to do this is to work by box number. Whenever an in-process check is performed, note the box number of the last box produced. It is important that all results are recorded.

In-process check sheets should be used (Figure 3.2 shows a possible layout for an appearance check sheet). These can be copied and sent to the pharmaceutical company as a protocol. Once the reject

COMPONENT DESCRIPTION		COMPONENT CODE	CUSTOMER CODE	BATCH NUMBER	DATE PRODUCED
PHARMACEUTICAL COMPANY BATCH NUMBER (ALLOCATED WHEN RECEIVED)					

Time	Box Number	Number of Samples	Appearance Defects (State Number and Type of Defect If None – State None)		Initials
			Major A.Q.L........	Minor A.Q.L........	
			If number of defectives is outside the acceptance limits, state rejected and boxes X to Y removed from the batch.		

COMMENTS

QUALITY ASSURANCE SIGNATURE	DATE

Figure 3.2. Appearance in-process check sheet.

material has been identified, it must be completely removed from the production equipment and placed in a quarantine area. This will ensure that it does not become mixed with the good material by mistake. The reject material can be 100% inspected to remove the defects, but **must** be sampled as per the statistical sampling method and examined for defects after the 100% inspection. This is because defects can be missed during an inspection and the efficiency may be very low in some instances.

Dimensional checks do not usually require taking large samples, as, for appearance. It is mainly at the machine set-up stage that dimensions have to be carefully monitored. In some instances, e.g., molded components, the dimensions may not vary over many weeks if the molds only wear slowly. These will require a periodic check of samples from each cavity of the mold. With such items as tubular vials and ampules, the dimensions will depend on how well the machine has been set up and may vary during a production run. Where dimensions need to be monitored in process, the sampling must be related to the number of forming heads or stations. The in-process sampling will usually require the monitoring of critical dimensions, noncritical dimensions being monitored only at the set-up stage.

All critical dimensions must comply with the limits specified on the drawing. Noncompliance with the limits, i.e., rejects, must result in the isolation of all items produced since the last in-process check. Again, as with unacceptable appearance, these must be completely removed from the production equipment and placed in a quarantine area. Unless go, no-go gauges can be used, it may not be possible to sort into acceptable or reject material, and the isolated items must be rejected and destroyed.

Any other parameters considered to require in-process monitoring must comply with the specification limits, e.g., component compatibility, physical parameters of rubber or plastic components, hydrolytic resistance testing of glass vials, etc.

5. Quality Assurance Involvement

It is not considered necessary for quality assurance to carry out the in-process testing. Production staff, and in particular the machine operators (wherever possible), should perform the testing, although training must be given by quality assurance staff. This will ensure that the production operators are responsible for the quality of production, rather than the quality assurance staff. Quality assurance staff are there to monitor the quality of production and to ensure that standards do not vary. It is only necessary for quality assurance to take the occasional random in-process sample during a production run. At the end of a batch, quality assurance will check the in-process results, and if satisfactory, will immediately pass the batch for dispatch to the customer. A copy of the in-process control results will then be sent directly to the relevant quality assurance representative in the pharmaceutical company as a protocol. Sending the protocols with the batch may result in their getting lost.

C. Production Operation Security

The most serious defects that can occur are rogue components, incorrectly printed components and incorrect material of construction for primary components. They can all have a potentially serious effect on the customer. Hence, any one instance of the above problems going undetected is likely to result in a batch recall from the market; therefore they must be avoided. The security precautions necessary vary with each defect, therefore they will be considered individually.

1. Rogue Components

These can be either rogue printed items or rogue primary items made from a different material than specified by the pharmaceutical company. If such defects are present at very low levels, there will be a high risk of them not being detected, especially if they are the same size, shape, or print color. A rogue printed item may state a different product name or different strength. Such errors could lead to the death of a patient; therefore, even **one** such item is unacceptable. With a primary component made from the wrong material, degradation of the pharmaceutical product may occur. This would render the product ineffective.

To avoid rogues during component manufacture requires close control of reconciliation, machine clearance, and segregation of each order produced.

a. Reconciliation.   The quantity of mix or components received onto a machine must be known. On completion of the operation, the components prepared must be counted. Variations in the final count against the expected output or quantity received must agree within preset limits. Counts outside the limits must be investigated. One hundred percent reconciliation usually cannot be obtained. Therefore the odd rogue item will not be detected, only significant quantities of components either going astray or being included, by error, in the batch will be detected.

b. Machine Clearance.   This ensures that the machine or production line is thoroughly checked for rogue items after completing an order. To do this correctly requires well-trained staff. Every part of the machine capable of harboring a rogue item must be identified. It is suggested that a diagram showing all of the areas to be checked is attached to the machine for the operator to refer to. This will be particularly useful for training new staff.

It is also important to check the area surrounding the machine/ production line for rogue items. The checking of the machine and area is a more effective way of ensuring that rogues are detected than relying on reconciliation.

When molding primary components, it is important to ensure that the machine is thoroughly cleaned prior to molding with a different material. Ideally, a defined amount of components should be rejected after start-up (the quantity being based on experience of the amount of material required to flush the machine, but initially being determined by analysis). A good indication can be obtained by molding with a white material after a highly colored material (e.g., dark blue) and monitoring the color of the white components.

c. Segregation of Each Order Produced.   This relates to the general security needed throughout the factory. Each production area must have clearly defined boundaries where neither unauthorized staff can enter, nor components produced on other machines can be placed. The components produced for a particular order must be placed into clearly labelled containers. The container labels must be produced in a secure manner, checked, and 100% reconciled. (Computer label generation is probably the best method.) A rogue label placed on a container could mean a complete container of incorrect components being sent to a customer.

Once the completed order is moved from the production area to the stores and awaits dispatch, the order should still keep its identity, that is, it should not be placed on a pallet with another order intended for a different customer.

2. Incorrectly Printed Components

The security concerning the print on the components depends on the following:

1. Artwork being correct
2. Preparation of the printing plates, step, and print operations
3. Correct printing plates placed on the machine that are not damaged

a. Artwork. This should be supplied by the pharmaceutical company requiring the printed components. Hence the component manufacturer must rely on the pharmaceutical company to ensure that the text is correct on the artwork. It is likely that the print text will be changed occasionally by the pharmaceutical company. This requires several actions if errors are to be avoided.

These are

1. The pharmaceutical company will send new artwork, giving clear instructions as to when the new text is required on the components.
2. Orders placed with the supplier prior to the text change but that have not yet been manufactured must be considered. The decision as to whether these need to be produced using the new artwork must be made.
3. The supplier must destroy the old printing plates to ensure that they cannot be used in error.
4. The pharmaceutical company's packaging materials laboratory needs to be aware of exactly when components are to be printed using the new text to ensure that the correct standards are in use.

b. Preparation of the Printing Plates. Clearly defined instructions, well-trained staff, and 100% checking that the print is correct are all necessary. The easy way to check printing plates is to make a print. This will ensure not only that the print is correct, but that an even impression is obtained. If multiple printing is to be done, for example, with labels, each printing must be checked. Gang printing must not be done when preparing labels, cartons, and leaflets for the pharmaceutical industry, i.e., printing items with a different text at the same time. The security risk is too high in that the separating of differently printed items is very difficult to control.

An easy way to check print (particularly for items where there is a large amount of text) is to use an image comparator. This relies on having a standard item that is **known** to be correct.

c. Correct Printing Plates Placed on the Machine. Each printing plate must be clearly identified and stored in a secure place, with authorized access only. When required for use, each plate must be carefully examined to ensure that there is no damage. Once set up on the machine, the "first-off" sample must be carefully examined to ensure that the print and color are correct.

3. Incorrect Materials of Construction for Primary Components

The main areas for consideration are the preparation areas for rubber and polymer mixes. This requires ensuring that the correct raw materials are used and that they are mixed in the correct quantities.

a. Raw Materials. The component manufacturer must ensure that all raw materials are adequately labelled and that a certificate of conformance is supplied with each batch received. If laboratory facilities are available, each batch received should be identified and examined for appearance.

It is advisable to examine the raw material manufacturer's premises to ensure that their manufacturing and control systems are adequate.

b. Preparation of Rubber and Polymer Mixes. The following rules must be adhered to if mistakes during the manufacture of a polymer mix are to be avoided. Either the wrong ingredients or even the incorrect quantities of ingredients could cause degradation of the pharmaceutical product for which the molded components are to be used. Such mistakes could also alter the physical characteristics of the polymer mix, which may affect the function of the component.

Each mix must be prepared separately, with no other raw materials than those required for the mix being produced in the area.

Clearly defined manufacturing instructions must be available and followed.

The batch number and quantity of each raw material weighed must be recorded on a manufacturing log sheet. The containers into which they are weighed must be clearly labelled with the material description, batch number, and quantity.

Weighing scales used must be kept regularly calibrated and records of the calibration kept.

Mixing times must be validated to ensure that a homogeneous mix is obtained. The materials must be mixed for the exact time proved to be necessary by the validation.

The final mix must be placed in a clearly labelled container, stating the mix reference number, the component for which it is to be used, the batch number, and the pharmaceutical customer. This is to ensure that the mix can be readily identified when taken to the molding machine.

D. Good Housekeeping

It is always pleasing to the eye to see a clean and tidy factory with well laid-out equipment. Visitors will be immediately impressed (first impressions mean a lot), and they will think the factory is operated in an efficient manner. An untidy factory will give the opposite impression.

Good housekeeping applies to the whole factory. It concerns not only keeping all areas clean and tidy, but also the layout of equipment to make the best use of the space available.

Many factories have a space-shortage problem, therefore examining the flow of materials through the factory and organizing the raw material storage, production equipment, and finished component storage to minimize materials movement will be an advantage.

Ensuring that staff are disciplined to keep their areas tidy and having clearly defined areas for placing items and materials in each area will enable the factory to be run more efficiently. This will also reduce the risk of mistakes and improve security. An example of what should not be done is the random placing of the loaded pallets in walkways, etc.

Storage areas can often be the most untidy. A stores location system should be used, with each shelf clearly marked. All items must be immediately placed on a shelf and not left in an aisle. Such obstructions can significantly delay the location and removal of items from the stores. A raw materials stock control system should be used to ensure that stock rotation occurs (i.e., oldest material is used first) and that replacement stocks are ordered in sufficient time to prevent delays in production. Also, monitoring to ensure that excessive stocks are not kept and that redundant materials are removed from the stores and not kept forever more in a corner gathering dust is necessary.

E. Tooling Control

It is very important for a molded component supplier to ensure that molds are made correctly. This is the starting point for the manufacture of a component, and if sufficient care is taken at this stage, then component dimensional problems can be avoided.

1. Tooling Manufacturer

The tooling manufacturer must be supplied with the following:

1. A detailed drawing showing all the component dimensions, indicating which are critical. This drawing will be different to that supplied by the pharmaceutical company because glass, rubber, and plastic components will shrink after molding. Therefore the drawing will be made for a mold that is larger than the component size required.
2. The techniques to be used for measuring each cavity of the mold must be defined.
3. The type of metal to be used for making the mold and

whether coating is required. A tool that wears slowly is advantageous to the component supplier and pharmaceutical company in that fewer dimensional checks will be required. Molding with polyvinyl chloride (PVC) is very corrosive, therefore, a surface treatment to prolong the mold life is required. The surface treatment should only be done once the mold dimensions have been checked.

The tooling manufacturer must check the dimensions of all the cavities in a mold, even if there are hundreds of cavities. This may take several days, but it will be worth it. Particular attention must be given to measuring the critical dimensions, especially if they are difficult to measure in the molded component, such as soft-rubber components. For example, to measure the depth of flange in the mold for a rubber plug is easy using a depth gauge (unless it is molded in a web form, where the web is the total or part of the flange thickness). It is important to check the flange depth in several positions for each cavity to ensure that there is no variation in depth.

2. Commissioning the Tooling

The tooling should be commissioned before starting production to ensure that each cavity in the mold is producing components to the correct dimensions. If the mold is to be treated, i.e., surface hardened or coated, then these commissioning trials should be performed prior to treatment. This is because if machining is necessary, then this could not be done after treatment. The molding should be done using the formulation mix required by the pharmaceutical company.

Each sample should be checked for dimensions using exactly the same techniques as agreed upon with the pharmaceutical company (see Chapter 1, Section III.C.1.b). This serves two useful purposes in that it gives a check on the tooling manufacturers measurements and confirms the rate of shrinkage expected. Once all the cavities have been found to be dimensionally correct, the tooling can be treated (if required) and put into production.

When replacing or refurbishing a mold, the relevant cavities must be checked for dimensions as well as the first molded components. The pharmaceutical company must be informed when a mold is refurbished or renewed, and samples from each cavity should be sent for them to double check the dimensions.

One practice operated by many suppliers of molded components is to make a mold to give component dimensions on the minimum of the specification (rather than the midpoint of the limits). This increases the life of the mold, because it will gradually wear to give component dimensions that meet the top specification limits. Considering the high cost of such tooling (particularly hot-runner tooling), this is an understandable practice. This increases the importance of ensuring that the pharmaceutical company cross-checks the component dimensions, since the precision and accuracy of the measuring techniques used might lead to arguments as to whether the components meet the specification. The best action from the component manufacturer's point of view is to aim for dimensions that are just above the minimum limits by an amount that takes into account the variations likely to occur, i.e., determined by the precision and accuracy.

F. Standards in the Manufacturing Premises

The general standard of premises for manufacturing primary components is usually less than that in a pharmaceutical factory, probably because the supplier is making materials for a less strict industry that takes larger quantities.

A rubber manufacturing company preparing rubber plugs (for vials) ideally should mold the components in a clean room to the same standard as the pharmaceutical company preparation area. (See Chapter 6, Section IV). In practice, it may not be a practical proposition to construct a clean room. Therefore, an alternative is to have localized laminar flow protection of the molded components. This would be required at the molding machine take-off point. Ensuring that the components are collected in a fiber-free container and sealed prior to removal from the localized laminar flow will prevent component contamination. The staff sampling the components for in-process checks and sealing the containers should wear nonshedding protective clothing that is tight fitting around the neck and sleeves. Hair must be completely covered by suitable headwear.

Operating to such standards may be a significant advantage to the pharmaceutical company, because the components may not require washing prior to use. It is better to ensure that contamination does not occur rather than to rely on a washing system to remove the contaminants in the pharmaceutical premises.

III. COMPONENT SPECIFICATION

It would be a waste of time to resolve the quality problems and then find out that the supplier cannot meet the specification. Therefore, all aspects of this must be discussed in detail. Also, appearance standards must be given to the supplier, showing examples of major and minor defects. An exact duplicate set of standards must be kept by the packaging and testing laboratory of the pharmaceutical company. These are essential if the same standards are to be applied when staff changes occur.

It is likely that new defects will be identified in time, and these must be classified and added to the standard charts. The most contentious aspect of appearance standards is the fine dividing line between major and minor defects. Therefore, this borderline area must be well represented in the standards and not just the **definite** major and minor defects.

Security of the appearance standards is very important, that is, there must be no possibility of their becoming mixed with a batch of components being produced.

These standards must therefore be securely fastened to a chart and clearly labelled. Ideally photographs of the defects, whenever possible, rather than actual samples, would be more secure.

IV. COMMUNICATION OF
QUALITY PROBLEMS

When buying millions of components from a supplier, it is inevitable that there will be the occasional quality problem when using the components. Therefore, rapid communication of all the details of the problems observed is necessary if corrective measures are to be taken

		PRODUCT REFERENCE

COMPONENT DESCRIPTION	CODE NUMBER	BATCH NUMBER

DATE RECEIVED	SUPPLIER BATCH NUMBER	SUPPLIER	DATE OF PROBLEM

PRODUCT NAME	CODE NUMBER	BATCH NUMBER

PROBLEM DETAILS

    (Include details of where the problem was found, effect on production, number and percentage of defective components found, whether major or minor and the decision made).

NUMBER OF SAMPLES INCLUDED (If None - State None)

SIGNATURE OF QUALITY ASSURANCE REPRESENTATIVE	DATE

Figure 3.3. Problem notification.

quickly. This is best done by telephone, by the quality assurance representatives, followed by written details and samples being sent to the supplier. A simple form can be designed to communicate these details (see Figure 3.3). A record of all such complaints must be kept by the pharmaceutical company.

There may also be times when a supplier has a problem that has not been met before, e.g., noncritical dimensions outside specification, or a new appearance defect. Rather than making a decision in isolation, i.e., either rejecting materials or deciding to take a chance and send them to the customer, communicating with the quality assurance representative of the pharmaceutical company will ensure that the right decision is made. This is better than a rejection by the customer and the resultant inconvenience and delay in having to supply another consignment of components.

V.   SUMMARY

Hopefully, the aspects covered in this chapter have shown how attention to detail and continual communication are absolutely essential if misunderstandings and rejections are to be avoided. In addition, a clearly defined mechanism for follow-up communication will ensure that the system will work.

4

The Packaging Material Quality Control Laboratory on the Pharmaceutical Premises

I. INTRODUCTION

The aim of packaging material quality control is to ensure the quality is right at the time of manufacture, hence eliminating testing on the pharmaceutical company premises. There will still be a necessity to periodically monitor components received, particularly with new suppliers, problem suppliers, or new components. Investigating packaging problems and customer complaints will also need laboratory facilities. (The same laboratory could be used for testing all raw materials received into the factory.) Therefore a suitable laboratory layout and testing procedure will be required.

II. LABORATORY CONSIDERATION

Considerable thought must be given to the laboratory location and layout to ensure that efficiency and security are maintained. The main points to consider are

Receipt and storage of samples

Inspection facilities

Chemical testing

Measuring facilities

Return of components to stores (when full boxes have been obtained for sample removal)

Recording and storage of results

Reconciliation and security of components

A. Location

There is only one suitable place for a testing laboratory to be located—incorporated in the warehouse receiving the packaging materials.

This will have several advantages:

1. Rapid access by the laboratory staff for sampling of component batches
2. Improved communication between stores and laboratory personnel concerning procedures and problems
3. Improved security and reconciliation when full boxes of components have to be transported to the laboratory for statistical sampling and then returned to stores
4. Rapid sentencing of batches on completion of testing

A suitable laboratory location is shown in Figure 4.1.

This enables a smooth flow of materials through the area from receipt, through to the packaging and dispatch of the product.

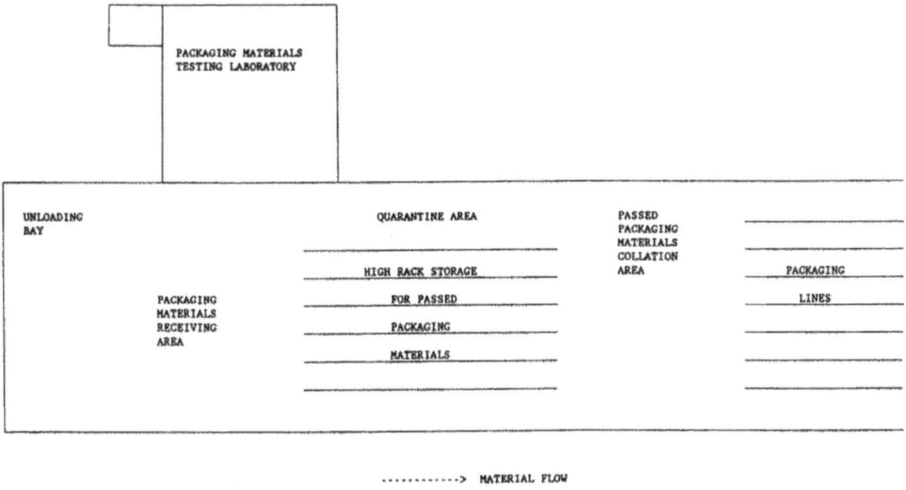

............> MATERIAL FLOW

Figure 4.1.   Suitable laboratory location.

B.   Layout

A suitable basic laboratory layout is shown in Figure 4.2. The area should be constructed with smooth walls and a coved floor for easy cleaning. There should be localized laminar flow protection to class 100 for sampling and inspecting primary packaging materials. This is to reduce the risk of particulate contamination of primary components during sampling and testing. The standard is the same as the component preparation area (see Chapter 6, Section IV). Localized laminar flow protection is required because of the large quantitites of fiber-shedding packaging materials in the area, such as cartons. The alternative is to have a completely separate area for primary component inspection. There must be regular cleaning schedules for the area.

1. Entrance and Changing Room

Staff working in the laboratory should change into nonshedding protective clothing before entering the working area. A one-piece garment should be worn that is fitted around the neck, ankles, and wrists. Hair must be completely covered by suitable headwear.

2. Samplers Working Area

Here the sampling staff will bring the complete pallets of material for either removing samples from the required number of boxes in the batch or for removing complete boxes from the batch. Once a batch has been sampled, the consignment will be placed into a quarantine area to await sentencing. All sampling equipment will be stored in this area, i.e., sampling bags, "sampled" labels, tape for resealing boxes, etc. A sink is necessary because the samplers will need to wash their hands regularly throughout the day.

3. Storage Area

Samples awaiting testing will be clearly labelled and placed in the storage area. Whether racking is used or a combination of open space and racking will depend on the quantities and types of components received. When large quantities or large items are received, then storing them on pallets while awaiting inspection may be more practical. Any pallets taken into the laboratory should be clean and nonshedding and constructed of plastic, aluminum, or stainless steel.

4. Inspection Area

It is suggested that this be one large, open area with movable barriers separating each technician. This will allow the layout to be changed quickly and easily, depending on the circumstances. If the laboratory is facing north as shown (with windows), then technicians examining colored components would be best situated along this wall. Daylight from the north (in the northern hemisphere) is the best lighting for checking color against a standard. The area allocated to each technician should be sufficient to allow the samples of the batch under test to be stored without overflow into another technician's area.

5. Chemical Testing Laboratory

This area should have all the equipment necessary to perform any testing required on packaging components and will include facilities for:

1. Hydrolytic resistance testing of glass—small autoclave
2. Ash determinations of plastic components—fume cupboard

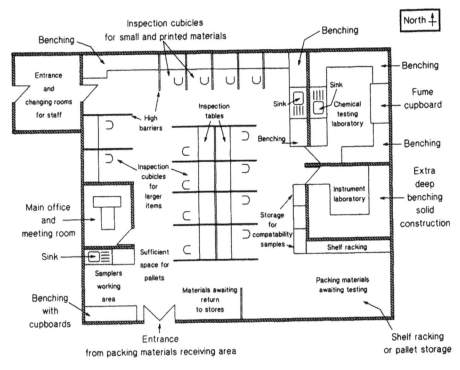

Figure 4.2. Basic laboratory layout.

3. Infrared testing equipment for monitoring adhesives and plastic components—infrared spectrophotometer
4. UV testing equipment for measuring irradiation levels in dosimeters—UV spectrophotometer

6. Instrument Laboratory

All sensitive measuring equipment should be set up and used in this area. Such equipment needs standing on very firm benching to minimize vibration, which can affect the equipment operation. The benching may need to be extra deep for such items as optical projectors. Other pieces of equipment likely to be required would be callipers, micrometers (including computerized calliper and micrometer systems), dial gauges, and balances. The standard weights and gauging blocks should also be kept in a locked cupboard in this area, with access only by trained technicians.

Each piece of equipment must have a reference number, which must be quoted in the relevant workbook whenever used. The equipment should be calibrated regularly if in regular use and zeroed whenever used. A record of all data must be kept relating to the specific equipment reference number, and the standard used must also be recorded.

A certificate of calibration must be obtained with all standard weight and gauging blocks each time they are calibrated. Recalibration must be done regularly, the frequency depending on how often they are used.

7. Main Office and Meeting Room

The head of the packaging materials testing laboratory should have an office along with a facility for having meetings with supplier representatives. The laboratory supervisor(s) should have an office in the main inspection area to ensure close contact with the technicians.

8. General Comments

The laboratory is shown as facing north, which is ideal for examining the color of cartons, labels, etc. This area should be used for examining these items.

Cupboards are necessary for storing components for compatibility and other testing. If many different types and sizes of bottles and vials are received, then samples of every type of cap, nozzle, or seal are required for compatibility checks. Also, samples of every type of bottle for checking the caps and seals are necessary.

III. RECEIPT, SAMPLING, TESTING, AND SENTENCING OF PACKAGING MATERIALS

A. Receipt

When a consignment of components is received, the storeman must check to see the consignment is clean and that there is no transport damage. Then he should make sure that the correct quantity of components have been delivered and that each container is correctly labelled. The required quantitity of identity labels should then be generated and attached to each container in the consignment, next to the manufacturer's label (for easy cross-checking of data). The identity label should state the component name, code number, batch number, supplier, date received, label number, and the words *under test* or a similar statement if a manual sentencing system is to be used. The best way to produce these labels is via a computer label-generating system. Then the exact number of labels required can be generated, minimizing label security problems. This can be part of the main factory computer system, which controls component stocks and usage. Each consignment is placed on a different pallet, as mixing components on one pallet can lead to mislabelling.

B. Sampling

The sampler should check the total number of containers and that the manufacturer's and storeroom's labels are correct. Then the $\sqrt{n} + 1$ of the number of containers should be removed from the consignment for sampling. These should be transferred to the sampling area of the laboratory if samples are to be removed, or else placed on the storage shelves if the technicians are to do the sampling. Dusty containers should be cleaned prior to taking them into the laboratory. Provided that the supplier has either placed a dust cover over or shrinkwrapped each pallet, this should not be a problem. When the samples are removed, the container must then be resealed and a label with the word *sampled* on it must be attached, stating the quantity removed, the date sampled, and the sampler's name. Each consignment of components awaiting testing and sentencing should be placed in a quarantine area. Dust covers should be placed over each pallet of components after sampling.

C. Testing

Before the testing of components can begin, it is necessary to have component specifications (including drawings, artwork, and testing methods, as detailed in chapter 1) and trained staff.

Each technician should have a clearly defined area for inspecting components. This is treated as a production operation, that is, only one batch is allowed in the area at a time. All components are reconciled into and out of the area. Once examination and testing is complete, the work area is examined for rogues before another batch is started. No unauthorized staff should be allowed in an inspection area when a batch is being examined. The reasons for the strict security are as follows:

1. If the samples are to be returned to the main batch, there must be no risk of rogue components being introduced during inspection.
2. If a rogue component was found during inspection, there must be no doubt that the rogue was in the batch as received and not introduced by error in the laboratory. To wrongly accuse a supplier of introducing rogues would create many problems.

When components are being examined for defects, either daylight (ideally north light) or a filament lamp should be used. (Fluorescent light is not as good for seeing defects or checking colors.)

Samples required for dimensional checks or chemical testing can be taken during the visual inspection either at random, by cavity number, or by die station, depending on the component being tested. Once removed from the batch, they must be identified by the batch number and the tests to be performed. (Placing in a labelled plastic bag is an ideal method for such items as caps, nozzles, rubber plugs, ampules, vials, etc.) It is sometimes viable to test several batches at a time to save time and manpower. An example may be determining the hydrolytic resistance of glass vials or ampules. It is suggested that clearly identified samples be placed in a secure place while awaiting testing. It cannot be overemphasized how important it is to ensure that batch security is maintained during testing. Common areas, such as the chemical testing and instrument laboratories, are high risk areas. This is because it is almost certain that more than one technician will be working in these areas at a time, with several batches of components.

All results relating to a batch of components must be recorded in a workbook (this could be computer generated, as mentioned in Chapter 1). The workbook is the main link document between the supplier, packaging component, and pharmaceutical company component batch number. Without this it would not be possible to trace the batch of components back to the supplier. The following data should be included in the workbook:

1. Component description
2. Supplier's name
3. Supplier batch number
4. Component code
5. Pharmaceutical component batch number
6. Date received
7. Supplier's protocols
8. Component testing results, including signature(s) of technicians and date tested
9. *Use before date* may need to be stated for some components due to deterioration, for example, prefolding in cartons on adhesive and self-adhesive labels
10. Date sentenced and authorizing signature

An example of a suitable workbook format cover is shown in Figure 4.3. This is considered to be a prime document and must be kept along with the product manufacturing records.

D. Sentencing

With consignments of components received that are considered suitable for immediate use without testing, it may be advisable to place an *under test* label on these, to complete a workbook front cover, and to sentence them as *passed* when the supplier protocols have been examined. This will serve several purposes. Firstly this workbook will ensure that there is a tie-up between the supplier and pharmaceutical batch numbers for tracing consignments used with a product. Secondly, it ensures that the in-process protocols have been received, checked by quality assurance, and attached to the workbook. Thirdly, it ensures that errors do not occur in the receiving area. The storeman will have to label all consignments of components received in exactly the same way, and not label some batches with *under test* labels and others with *identity* labels. One cumbersome aspect of this method of labelling consignments that do not require testing is that *pass* labels will have to be attached over the *under test* labels. A computer system could control this easily, but the in-process protocols still have to be examined and cross-related to the pharmaceutical batch number allocated to each consignment of components.

Components that have a definite life must have the *use before date* printed on the pass label.

Batches of components rejected must be labelled with *reject* labels (preferably printed in red) and removed from the quarantine area as quickly as possible, that is, destroyed or returned to the supplier.

See Figure 4.4 for the sequence of events from the receipt of packaging materials until sentencing.

IV. DOCUMENTATION

A. Testing Methods

The methods to be used by the laboratory technicians (which will be identical to those used by the suppliers) must be clear and concise, but should include every detail of the test and the techniques to be used, including diagrams whenever possible. Measuring techniques are very important, as indicated in Chapter 1, Section III. If such important aspects are left open to interpretation by the technician through poorly written methods, then incorrect results will be obtained. The methods are training documents.

B. Method Layout and Security

A procedure for writing methods should be produced that clearly defines the format, numbering system, updating, and distribution. Each method should have a specific reference number and version. The version number is changed with each method update. Whenever an updated version is issued, all the old methods are withdrawn and destroyed. Therefore it is essential to keep a list of names of everyone to whom the method was issued. The master copy is stored in a secure

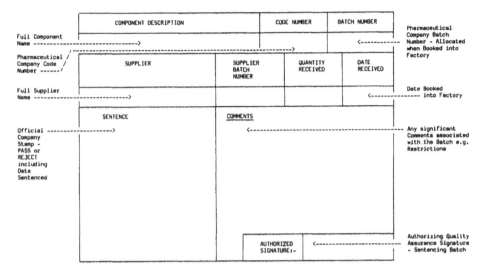

Figure 4.3. Workbook front cover.

place and copies are only issued from the master, recording the name of the person to whom it was issued. Unauthorized photocopying of issued methods must not be allowed, otherwise it will be difficult to locate, recall, and destroy old versions when a reissue is necessary. Each page of the master copy of the method should have an authorized signature to prevent unauthorized changes in the method. Ideally, each page should state *written by, approved by,* and *authorized by,* along with names and signatures.

V. INVESTIGATING COMPONENT PROBLEMS OCCURRING WITHIN THE PHARMACEUTICAL PREMISES

It is inevitable that problems in the usage of components will occur, even though a consignment has been sampled and tested. Therefore there must be trained investigators within the packaging-materials quality control laboratory who will investigate such problems in the production areas.

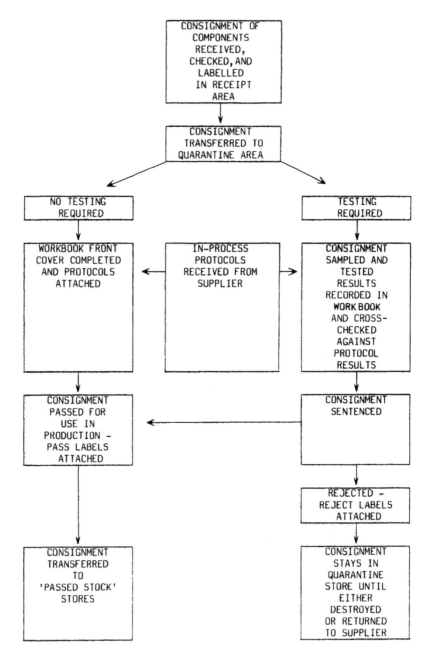

Figure 4.4. Sequence of events from the receipt of packaging materials until sentencing.

An investigator should ideally have the following abilities and training:

- be a practical person with the ability to watch a machine in operation and be able to understand how it works
- be capable of making a logical and common sense assessment when a new situation or problem is encountered
- have an understanding of both production and engineering aspects concerning the various production operations, as well as extensive quality assurance experience
- have an analytical mind with the ability to ask the right questions, assess all the data obtained, and make a rapid decision
- be fully aware of his or her own limitations with respect to making quality-related decisions

When the investigator arrives at the production line, it is essential to talk to the supervisor and staff involved with the problem and to find out how the problem was discovered and the exact nature of the problem, and to obtain samples of both good and bad components. If possible, ask to watch the problem being generated. This will enable an independent assessment and will eliminate unfounded opinions, e.g., an engineer blaming components because a machine problem cannot be identified. A quick visual comparison of the good and bad components can quite often determine whether there is a component problem. It may be necessary to have the dimensions checked or other tests carried out in the laboratory. This will take time and a filling or packaging line cannot be out of production for long periods when 10–15 staff are involved. In addition, this would not solve an ''out-of-production'' situation. Whether the components are at fault (i.e., out of specification) or not does not resolve the immediate problem.

If a component problem is definitely identified by a visual examination, then the actions to be taken are shown in Figure 4.5, which details the ideal course of action to minimize final batch quality problems and to maximize output. This may not be possible if there are no components left in stores; in such instances either production ceases completely or the defective components are inspected prior to use. Inspection should only be used as a last resort and must be carefully monitored by production and quality assurance staff. Even after taking such precautions, there still is a risk of defective components appearing in the packaged product. In addition, this requires extra staffing and slows down production.

If the component problem is in a sterile area where the components require preparation prior to use, such as washing and sterilizing, the defective components should not be inspected. Human contact with sterile components must be minimized to prevent bacteriological contamination.

When it is not obvious that the components are at fault (without laboratory testing or measurement), a great deal of time and effort can quite often be saved by asking the following questions:

1. Have any machine modifications been made recently? This could affect the machine capability in handling the normal component variations.

2. Are there any new production staff operating the equipment? Inexperienced staff may not know the many minor adjustments and attention necessary to keep the equipment running.

3. Are there any new engineering staff setting up, maintaining, or repairing the equipment? If they have insufficient experience with the equipment, then faults may not be diagnosed and the blame may be wrongly placed on the components.

4. Has the machine operating speed been increased? This is likely to demand tighter dimensional limits on the components for smooth operation. Many problems may occur if the maximum recommended operating speed is exceeded.

5. Are there any changes to the product, such as different physical characteristics? For example, a cream with a lower density than normal might cause problems with obtaining the correct weight in a tube or jar without overfilling and could be mistaken for a dimensional problem with the container.

If none of the above questions resolve the problem, then testing or dimensional checks will have to be carried out on the components. If they are found to be out of specification, the batch of components

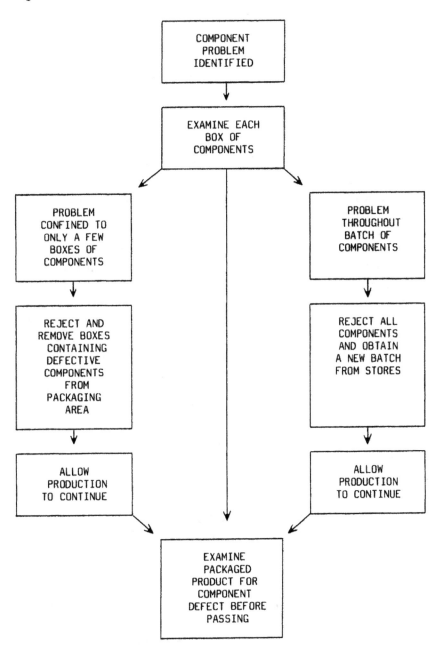

Figure 4.5.   Action taken when defective components found on a packaging line.

(including any remaining in stores) must be rejected. It would also be wise to check any later deliveries that are in the factory for the same defect.

Sometimes components that are within specification may not be usable on the production equipment because either a parameter not included in the specification was not thought to be critical or a parameter was incorrectly specified. Provided that the change necessary does not affect product quality, then negotiations with the supplier may be started to revise the specification. One point to note is that modified or new equipment may require completely different specifications, which may create problems, especially if the component supplier cannot meet the new requirements. The details regarding what must be considered when buying new equipment will be covered in Chapter 5.

It is essential that the details of the problem are recorded (whether associated with the packaging component or not) and stored with the relevant packaging materials workbook until after the end of life of the product. This is necessary in case a customer complaint is received. This record would enable an accurate assessment of the complaint and of the extent of the market problem. The form shown in Chapter 3 (Figure 3.3) should be used for detailing component-related problems. This form has a dual purpose, in that a copy is sent straight to the supplier as well as being included in the packaging material workbook. The form should be completed by the packaging component investigator. The production staff should also make an entry in an equipment log on the relevant packaging line relating to the problem, whether it is associated with the component or not.

Customer complaints will be considered in Chapter 7.

VI. STERILIZATION OF PRIMARY PACKAGING COMPONENTS BY GAMMA IRRADIATION

This has been considered as a separate topic because of the practical problems likely to be encountered from the initial preparations for irradiation through to transfer to the sterile area.

Gamma irradiation is probably the best method of sterilization for polymeric components because of softening and hence deformation by heat-sterilization methods. The minimum irradiation requirement is 2.5 Mrad (25 kGRAY).

A. Preparation for Irradiation

The components may either be packaged by the supplier and sent via the irradiation company to the pharmaceutical premises or packaged and dispatched from the pharmaceutical premises for irradiation.

If the supplier's molding machine, or at least the machine's outlet, is in a clean room, then the components can be packaged directly from the molding machine. If the bags are heat sealed as soon as each container is filled, then there will be minimal risk of contamination.

If the components have to be transferred to the pharmaceutical premises and then repackaged for irradiation, there will be excessive handling of the components, and hence a greater risk of particulate and bacteriological contamination. This is a considerable waste of time and an expensive method for obtaining irradiated components. It may even be necessary in the case of bottles to wash or blow them before they will be suitable for irradiation.

Packaging must be done in a clean environment, with the operators wearing protective clothing to prevent particulate and excessive bacteriological contamination. This should consist of a one-piece, non-shedding suit with tight cuffs (to prevent shedding out of the sleeves), complete hair covering, and surgical gloves.

The components should be double wrapped in polyethylene bags (or other sterility proof containers) of sufficient thickness to prevent puncturing during handling and transport. The bags should be heat sealed when full and not taped. (The bags used will themselves have to comply with a clearly defined specification regarding the thickness and leakproof seams.) Double wrapping enables the outer wrap to be removed immediately before transfer to the sterile area. The inner bag should be clearly labelled with the component description, component code number, batch number, quantity, and date packaged for irradiation and have an irradiation indicator attached. See Figure 4.6 for an example of a possible label format. A warning statement should be placed next to the indicator stating that the components must not be

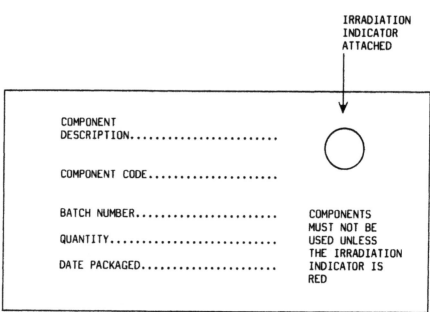

Figure 4.6.   Labelling of components for irradiation.

used unless the irradiation indicator has changed to the post-irradiation color.

The outer container should be nonshedding and should be of suitable dimensions to suit the irradiation company. It should be labelled in the same manner as the inner bag.

It is advisable to place dosimeters in the packaged components, for example, one box per pallet, attached by tape to an agreed place within the container (to prevent movement and damage during transportation). This will serve as a double check against the irradiation company checks (which must have at least two dosimeters in the irradiation chamber at a time). The container should be labelled *Dosimeter enclosed*. Whenever an irradiation company is to be used for the first time, the premises should be quality audited and then full validation studies should be performed for each component to be irradiated. This involves placing several duplicate sets of dosimeters throughout a pack (including the center of a pack). The position and quantities of dosimeters to be used will depend on the irradiation plant and should be decided after detailed discussions with the company. The irradiation company should conduct an identical validation study. The duplication of dosimeters in each position is advisable in case a rogue result is obtained. If two dosimeters give widely differing results, the experiment should be repeated. Single dosimeters giving rogue results always leave an element of uncertainty, in that it is difficult to know whether the fault lies with the irradiation process or the dosimeter.

B. Receipt of the Irradiated Components

Once the irradiated components have been received at the pharmaceutical premises, they should follow the same receiving procedure as all other components, although sampling is different.

The irradiated components cannot be sampled in the same manner as nonsterile components. Therefore, if samples are required it is suggested that they are prepared prior to dispatch for irradiation. Samples may be required for:

1. Preirradiation bacteriological check
2. Postirradiation sterility check—probably necessary if a high preirradiation bacteriological count is obtained
3. Testing to specification before irradiation
4. Testing to specification after irradiation (particularly for color, embrittlement, and possible dimensional changes)

The samples should be taken during component manufacture when in-process control samples are taken (this is the ideal method) or else sample from the $\sqrt{n} + 1$ of the finished batch of components. Pre- and postirradiation samples should be double wrapped in heat-sealed bags (in the same manner as batches to be irradiated). Preirradiation bacteriological check samples should be sent (together with any other nonirradiated samples required) directly to the pharmaceutical company. Postirradiation sterility check samples should be placed in a container of components for irradiation, together with any samples for other postirradiation tests. The container should be labelled *Samples enclosed.*

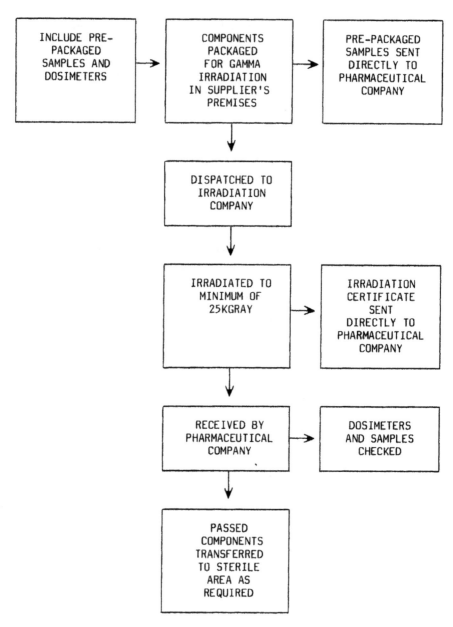

Figure 4.7. Flow of components sterilized by irradiation.

The irradiated samples can be removed by the samplers from the appropriately marked containers, along with the dosimeters.

A certificate of irradiation should be obtained with every batch irradiated. This should be cross-checked against the pharmaceutical company results and retained in the workbook.

C. Production Use of the Irradiated Components

There is a problem in linking a batch of components to the date irradiated, particularly since this date cannot be written on the inner bag until the components are in the sterile area. The irradiation date is stamped on the outer containers by the irradiation company. Therefore this date can be verbally communicated to the sterile area staff. This is not considered to be the most reliable method. Therefore, it is suggested that the date that the components were packaged for irradiation should be recorded in the workbook (obtained from the container labels). The same date should be noted on the filling documents when the components are used in the sterile area, and the component batch number should be recorded there as well. This will ensure the existence of a link to the date irradiated, particularly when the same batch of components is irradiated on more than one date.

The flow of components sterilized by irradiation is shown in Figure 4.7.

5

Packaging and Filling Equipment

I. INTRODUCTION

The importance of buying the correct equipment for a packaging function cannot be underestimated. The main problems that can occur from making the wrong decision are as follows:

1. The components currently available cannot be used efficiently without modifications to the component dimensions. A new specification would have to be agreed upon with a component supplier, and if these new requirements cannot be met, a serious problem is created. The preparation of equipment-specific component specifications is a negative step (costing time and money) when component rationalization should be the policy operated.
2. Quality problems created from the equipment not operating

correctly, leading to a high level of rejects, low productivity, and a possibility of poor-quality product on the market.

3. Expensive and time-consuming modifications to new equipment to try to remedy the above problems. This may only meet with partial success. An expensive machine then becomes even more expensive and may have to be replaced if a viable output of a good-quality product is to be obtained.

Therefore, before equipment is bought, a detailed buying specification is required, followed by extensive prepurchasing trials. The equipment should then be inspected prior to delivery, installed, commissioned, and validated before it is made available for production use. Figure 5.1 summarizes the sequence of events in buying packaging and filling equipment.

A flow chart needs to be prepared that shows the actions to be taken at each stage and clearly indicates the rate-controlling (critical path) stages of the project. This will allow a realistic timescale to be determined for completing the project, which can be used as the plan of action.

II. EQUIPMENT SPECIFICATION

A. Production Requirements

Before a specification can be prepared, a detailed analysis of the requirements is necessary. First, the production staff requiring the equipment must state exactly what they want the equipment to do, i.e.,

1. Component specifications. To be stated for all types and size or sizes of components to be handled.
2. Limitations. For example, state the volumes and limits of fill required for a liquid-filling machine (filling specification). When powder or granules are to be filled, state the range of particle size required to be filled, the density, and the moisture limits.
3. Filling or packaging rate.
4. Whether regular changes to different components sizes are

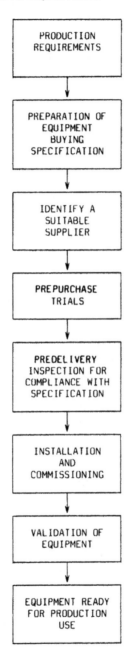

Figure 5.1. Sequence of events when buying packaging/filling equipment.

required (this is very important if the change-over takes a long time for a machine).

5. The location and environment in which the machine is to be used, e.g., sterile area, clean room, etc.
6. The filling or packaging-line layout if the equipment is to connect to other machinery. The infeed and outfeed positions on equipment forming a long packaging line need to be considered. This should include a layout drawing of the line.
7. Careful consideration by production staff is required to determine what monitoring gauges and devices are required to help ensure that product quality is maintained. These are classed as critical devices.

Examples are

1. Sparging gas pressure gauges (e.g., nitrogen) on an ampule- or vial-filling machine.
2. Temperature monitoring system for heat-sealing rollers of a tablet foiling machine.

When the requirements have been defined, the accuracy limits (or acceptable limits) must be determined for each device. The process requirements will decide what these limits should be. For instance, referring to example (2), above, there will be a temperature range over which a satisfactory foil seal will be obtained (in conjunction with the pressure and dwell time), which is determined during the pack validation. If the control required covers a range of $10°C$, then a temperature monitoring system accuracy of $\pm 2°C$ may be sufficient, provided that the operating limits are set at $\pm 8°C$, allowing for a possible $\pm 2°C$ inaccuracy. If a control of $\pm 1°C$ is required, then the monitoring system accuracy will need to be $\pm 0.1°C$, with operating limits of $\pm 0.9°C$. If the accuracy is $\pm 2°C$ with a required control of $\pm 1°C$, then it is inevitable that an ''out-of-control'' situation will occur. The importance of having a critical device with the required accuracy is summarized in Figure 5.2.

Once the requirements have been defined, they can be given to the relevant engineer to prepare the specification. It is advisable that

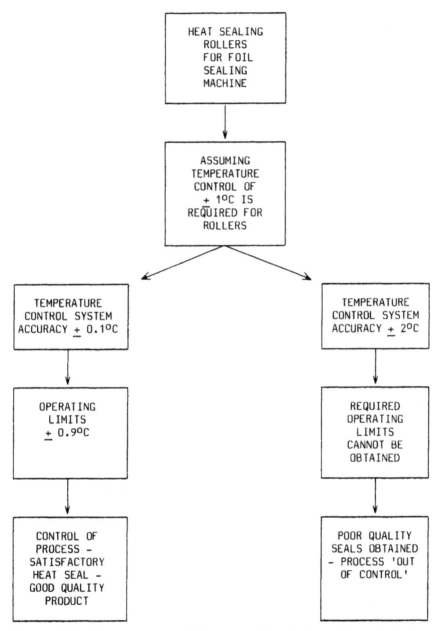

Figure 5.2. The importance of having a critical device with required accuracy.

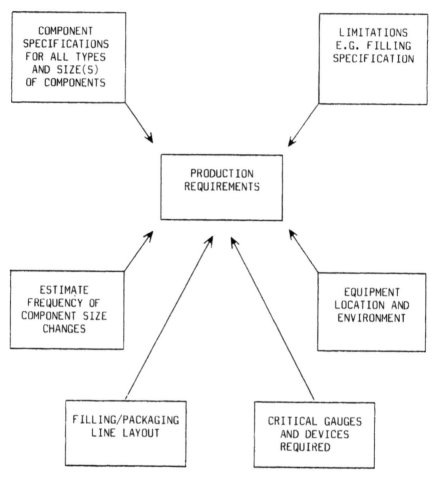

Figure 5.3. Buying new equipment—Production requirements.

the engineer be familiar with the type of equipment required. Figure 5.3 summarizes the production requirements.

B. Preparing the Specification

The specification should include the following:

1. Production Area Requirements

See Section A.

2. Services

State the services required and the standards to be used, e.g., electrical, plumbing, compressed air, etc.

3. Construction Material

It is not advisable to have painted equipment due to the risk of paint contamination. Stainless steel is the most suitable material for construction. State the standard of stainless steel that is required. Also state the standard of materials for the pipework to be installed, including the welds.

4. Basic GMP Design Requirements

Consideration must be given to safeguards to prevent products or components from becoming trapped in the equipment. Wherever possible buy equipment that will minimize this risk. A close examination of the equipment must be made to identify all potential risk areas and, if possible, modify to prevent entrapment. Packaging equipment should be skirted to the floor with stainless steel panelling to prevent rogue items getting underneath. It is better to design out the areas for rogue entrapment rather than have packaging machine operators search dozens of locations for rogues at the end of each batch produced, with inevitable missed rogue items due to human error.

Easy cleaning of equipment should be possible, particularly equipment to be used in a clean room or sterile area. The ability to easily completely dismantle filling equipment to ensure effective cleaning and sterilizing will help reduce the risk of bacteriological contamination and minimize the batch/product change-over time.

Easy access for maintenance and calibration is required. This is particularly important for sterile area equipment. A gauge that can be calibrated without removal from a main panel can save a lot of time and minimize the particulate contamination risk to a sterile environment. An example could be a pressure differential gauge on a laminar flow unit. Rather than having to dismantle the unit to access the gauge, have terminals fitted to the outside of the cabinet for calibrating purposes. Also gauges and electronic devices that can be easily removed if major time-consuming work is required (i.e., faults that cannot be easily detected) should be used. A replacement gauge or device could be fitted to maintain production. An example would be a defective missing label detector on a packaging line. It makes sense economically for the instrument technician to spend 2 hours in the workshop repairing the detector rather than 2 hours on the line, which would result in both lost production and the extra cost of the wages of the packaging line staff.

5. Lubrication

There is a potential risk of contamination of product with lubricants, particularly from any machine parts or gear boxes located above the product height, e.g., stirrer gear boxes. Therefore all oils and greases used must be FDA approved, edible lubricants. Many lubricants contain lead-based materials. When buying equipment, design features that minimize the potential risk from lubricants should be determined, although this may not always be possible. Where a potential risk cannot be avoided, extra safeguards can be incorporated, such as a cup arrangement on a stirrer shaft to stop oil running down the shaft. Usually a worn seal will leak slightly initially, and, provided the equipment is examined and monitored regularly, a potential product risk can be detected and hence corrected before a product is put at risk. An agreement on a suitable lubricant should be determined before buying the equipment. Ideally, the pharmaceutical company should have a list of the standard approved lubricants used for all equipment in the factory. This would eliminate the risk of a nonapproved lubricant being used by mistake. Equipment manufacturers should state which of these lubricants is to be used with their equipment.

6. Computer Control

If the equipment available is computer controlled, define exactly what type of computer system is required (try to standardize) and what security measures are required, for example, access codes, back-up systems, and alarms.

7. Standardization

Standardize whenever possible. For example, if the equipment is to include a filtration system, state the type, size, and grade that is currently used within the factory. This will alleviate problems with replacement stockholding.

Standardize the plumbing fittings and electrical switching systems. With small pieces of equipment, such as balances, try to buy one make only. This will mean that only one maintenance contract will be required.

8. Safety

Clearly define the equipment safety standards required by the factory (as well as national safety standards), for example, safety interlocks on all access doors to moving parts, and consider access for repairs and maintenance. Also take note of the placement of emergency stops on the equipment, and electrical and plumbing installation standards.

9. Critical and Noncritical Gauges and Devices

In addition to the critical gauges and devices stated under the production area requirements, there may be troubleshooting gauges required. These are classed as noncritical, in that they will not affect product quality if they are not working correctly, but may delay the repair time in the event of a breakdown. Deciding which gauges and devices are needed will require a great deal of understanding of the overall equipment and discussion with the manufacturers. The fitting of unnecessary gauges and devices that do not serve either production operation or troubleshooting requirements should be avoided. It is surprising to note how many pieces of equipment and plants have unnecessary gauges and devices fitted, which add an expense to both buying and sometimes installing.

Specifications should state that critical and noncritical devices must be sited to enable easy access for maintenance and calibration.

10. Service Manuals

State the number of service manuals required and what information they should contain, for example, full operating and maintenance instructions. Quite often these are unintelligible and hence of no value if the full requirements are not stated.

11. General Points

The final specification should be examined by a suitably experienced member of the quality assurance department, primarily to ensure that the GMP requirements are covered in the specification, such as

1. Material of construction, e.g., high-standard stainless steel if required for a sterile area
2. Provision for the necessary standard of in-line filters for services such as compressed air, nitrogen, etc, used by equipment where these gases will be in contact with the product.
3. Doublechecking to make sure that the filling and packaging specifications are correct
4. Consideration of the equipment design features relating to the cleandown and potential rogue problems

The above aspects must be checked by quality assurance for compliance to the specifications after installation (covered later in this chapter). The equipment specification requirements are summarized in Figure 5.4.

III. EQUIPMENT PURCHASE

Completed specifications can be presented to the relevant purchasing staff to identify possible suppliers of the equipment.

The next stage is for the equipment buyer and engineer to visit these potential suppliers. It is inevitable that any equipment identified is likely to require at least slight modification to comply with the specifications. Equipment requiring major modification should be avoided, since this could involve a prototype situation, with all the potential problems and expense that are usually encountered with new, untried equipment. It is always necessary to look for the new, latest state-of-the-art equipment if a company wants to continue to be competitive in the international market, although getting involved with a prototype machine will usually mean many months of problems and

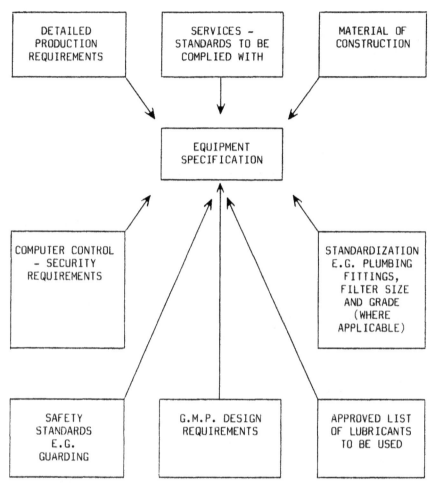

Figure 5.4. Equipment specification.

possible modifications before the equipment operates effectively. This situation is likely to result in low output, excessive product and/or component wastage, and many quality problems. Therefore, try to obtain the latest, but well-tried equipment. Preferably see not only the machine operating in the manufacturer's premises, but also in another customer's premises, if at all possible. Invaluable information can usually be obtained from talking to another customer.

One important point to look for in equipment suppliers is that they communicate and cooperate with component manufacturers. Machine manufacturers who fully appreciate the limitations and problems in making the components to be used on their machine are more likely to design the necessary flexibility into the equipment construction.

Too often the only real testing ground for the equipment and components is at the pharmaceutical premises.

A further point to look for in new equipment is simplicity. A complex machine will need more maintenance and may break down more often as it gets older, and therefore will require more engineering staff training than a simple machine. There will also be a need for extra spare parts to be kept on stock. Cleandown may be more difficult, with a greater risk of either contamination or rogues.

A. Prepurchase Trials

Once suitable equipment has been identified, packaging or filling trials should be organized. This cannot be expected to give any more than an indication as to whether the equipment is capable of meeting the specification requirements, mainly due to the practical problems of transporting sufficient trial material over a distance of perhaps thousands of miles. Such trials have to be performed under controlled conditions (even with placebo materials) to ensure that all materials are reconciled and destroyed after the trials. The pharmaceutical staff that need to be present at the trials are representatives of the production and engineering staff familiar with the operation of similar equipment.

The trials materials used should ideally cover the full dimensional range shown on the component specifications in order to allow a complete assessment of whether the equipment can cope with the variations to be expected over years of production. It is not practically possible for component manufacturers to produce such a wide range of components, therefore, another customer's experience (previously mentioned) can be invaluable. Unfortunately a competitor is unlikely to give access or details of such equipment. The alternative is to use components known to have the greatest dimensional variations that are available and to include samples from all component suppliers used (if single sourcing is not operated). Samples will probably need to be specially measured for the trials.

It is possible to get a great deal of information from the prepurchase trials by organizing the trials carefully. For example, suppose a cartoning machine is to be purchased that is capable of handling three sizes of cartons. Ensure that sufficient trial materials are available for at least 1 hour of operation at maximum running speed for each size of carton. Ask for the machine to be set up to run the smallest carton, run it at the slowest speed for a short time, and then run it at the maximum speed for the remainder of the trial (it is likely that more problems will arise at maximum speed.). Monitor any problems that occur and examine any components that give rise to problems (keep samples for dimensional checks if necessary). Then ask for the machine to be set to run the largest carton (the larger the size change, the more likely it is that readjustment will be necessary). Note the time it takes for the change-over, because this is an important downtime factor if regular change-overs will be necessary. Follow the same routine as for the small cartons. Finally, monitor middle-size carton trials. An analysis of the data obtained from problems encountered may indicate potentially serious problems. If the equipment is unable to cope with the normal variations expected on one of the carton dimensions, buying such equipment could result in many nonproductive hours.

During the equipment trials, monitor the amount of attention and staff required to run the machine. These are important for two principal reasons:

1. Equipment requiring a lot of staff will be expensive to run.
2. Equipment that requires 100% attention at all times will require very vigilant staff. It is difficult to find such dedication in staff, particularly on boring jobs. Therefore, equipment that jams up or goes wrong when operators take their eyes off of it for a short time should be avoided. Jam-ups inevitably lead to breakdowns; and then engineering staff will be required, quality control problems may occur, and there will be lost production time.

Carefully monitor the equipment set-up procedure carried out by the equipment manufacturer's engineer. If this requires many delicate adjustments, then problems may be encountered in setting up the equipment in the pharmaceutical premises.

The manufacturer will probably use their best engineers in a trial situation. A machine that can be set up without having to use a company's best engineer can again save a lot of downtime and quality control problems. The majority of rejects usually occur in the set-up stage.

Once a decision has been made on the equipment to be purchased, it is not advisable to tie a manufacturer to an unrealistic delivery date. A rushed job usually leads to mistakes. It is also unwise to move the equipment onto the pharmaceutical site if any modifications are required. Modifications are best done in the manufacturer's premises, where all the facilities are available to do a quicker and better job. The prepurchasing trial requirements are summarized in Figure 5.5.

B. Predelivery Inspection

Prior to delivery, the equipment must be examined by the pharmaceutical engineer to ensure compliance with the buying specifications and to ensure that any modifications requested have been made correctly. A simple way to ensure that all aspects of the buying specifications are correct is to have a column down the right-hand side of the specification, so that the entry for each part can be signed as correct and comments can be made as required.

IV. INSTALLATION OF NEW EQUIPMENT

Before the installation of the equipment, careful consideration of its positioning should be made for the following reasons:

1. Availability of services, e.g., compressed air, drains, the standards available, and the equipment requirements
2. Access for operation and servicing
3. Interoperation with other equipment to ensure streamlined operation

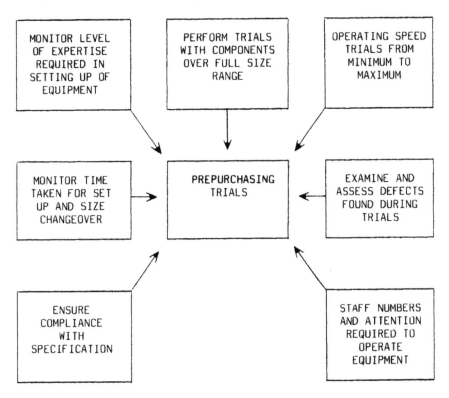

Figure 5.5. Buying new equipment—Prepurchasing trials.

A drawing showing the layout and services should be prepared, and a copy given to the installation engineers to minimize the risk of errors in machine placement. Figure 5.6 summarizes these preinstallation and commissioning considerations.

Once the equipment is installed, all aspects of the machine and its operations should be checked against the buying specification. This should involve examination by experts from all relevant areas of the factory, such as the safety officer, electricians, plumbers, quality assurance, and packaging development staff. Gauges and critical devices should be calibrated against certified calibrating equipment, which is traceable to the National Bureau of Standards. Records of each calibration should be kept, together with a copy of the calibration certif-

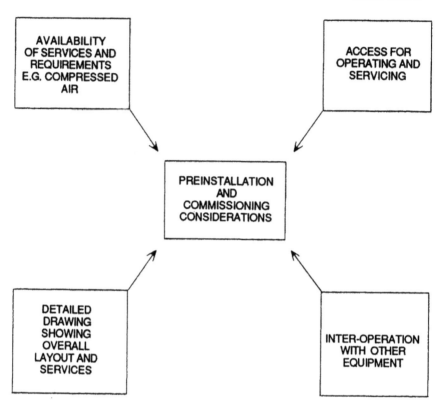

Figure 5.6. Buying new equipment—Preinstallation and Commissioning considerations.

icate. Each gauge should be clearly numbered and labelled with its monitoring function. All service pipes, e.g., compressed air, connected to the equipment should be labelled, stating what is being supplied, the supplying pressure, and the flow direction. The supply lines should also be labelled at their source for easy identification. For maintenance and safety reasons, electrical and electronic equipment wiring, and valves should be clearly labelled.

It is suggested that the equipment be identified by the pharmaceutical company for easy reference purposes, particularly with regard to maintenance and calibration. A simple form should be available for listing the equipment to be numbered and its location. This will provide a mechanism for ensuring that all relevant equipment is numbered. Figure 5.7 shows a sample equipment identification form.

Maintenance and calibration routines plus procedures should be prepared. These can be best controlled and monitored by using a computerized system. This will enable the relevant work to be scheduled and recorded, and will allow for a maintenance history to be built up. Planned maintenance can be either time or run-hour based. Run-hour mechanical maintenance is ideal, although it is difficult to plan in with production and will require run-hour meters attached to each machine, with a direct link to a central computer. Time-based maintenance is possibly better for electrical and electronic equipment, since time-related deterioration can occur even when the equipment is not in use. This is particularly important with earth continuity checks, which should be performed at least yearly.

The maintenance requirements can be split into three categories:

1. Essential quality-related maintenance, such as calibration of critical devices or sterile-area filter changes. This must be done when scheduled; otherwise the quality of the product will be at risk. The frequency must be based on the reliability of the equipment. Each maintenance routine prepared must be approved by the relevant production and quality assurance staff. The production person must ensure that the calibration limits for a critical device are within the process requirements. The quality assurance person must ensure that the routine content complies with the GMP requirements, such as reference to a specific calibration procedure, that calibrating equipment is stated, and that its calibration can be traced to the National Bureau of Standards.
2. Safety-related maintenance, such as electrical safety checks. These must also be done when scheduled, otherwise staff may be at risk.
3. General equipment maintenance, which should be based on what is essential maintenance to ensure that production is not significantly interrupted. This may not necessarily be exactly as the equipment manufacturer recommends. If the recommended maintenance frequency is based on a dirty environ-

FROM:-   (Name of Installation Engineer)

TO:-   (Name of Equipment Maintenance Engineer)

Would you please attach an Identification Number to the following
equipment:

REFERENCE NUMBER	EQUIPMENT DESCRIPTION (Include Serial Number)	LOCATION	ROOM NUMBER
(Number to be allocated by Plant Maintenance Engineer)			

Figure 5.7.   Equipment identification form.

ment, then an exceptionally clean pharmaceutical premises may warrant less frequent maintenance, for example, lubrication. It may be economically viable to consider breakdown maintenance, for example, with motors where either spares can be rapidly fitted or the old motor can be rewound within 24 hours.

It is not advisable to neglect preventative maintenance, since this will inevitably mean more breakdowns, excessive loss of production, a large team of engineering staff, and large quantities of spare parts. Although careful consideration as to the essential maintenance required must be assessed by experienced engineering staff, it is easy to do too much maintenance.

Planning for maintenance is not easy to do because scheduling maintenance in with the production schedule needs a good forward planning system for maintenance and a dedicated engineering resource. This means that the same staff cannot be used for equipment breakdowns. It may be worth considering doing off-hours maintenance, that is, either using night-shift teams or evening/weekend work.

Maintenance contracts by outside companies should be included in the maintenance system. These contracts should include a description of each piece of equipment to be maintained and details of the maintenance to be done. Recording the work completed should be done on factory computer-generated worksheets. This will ensure that adequate records of the maintenance work are available. The computer system then acts as a reminder of when a contractor visit is due, hence ensuring that maintenance is not missed.

The spare parts received with the equipment should be placed into the factory storeroom, code numbers allocated, and minimum/maximum order levels set. A simple form should also be designed to ensure that the spare parts become part of the factory system.

A summary of the installation requirements is shown in Figure 5.8.

V. COMMISSIONING OF NEW EQUIPMENT

Pharmaceutical engineering and production staff should be involved in equipment commissioning as part of their familiarization exercise. In

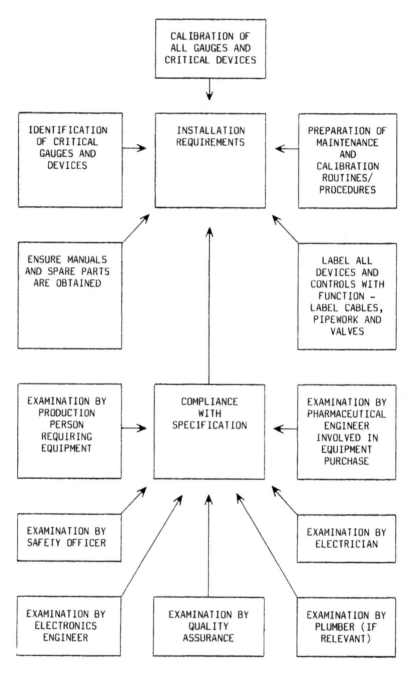

Figure 5.8. Buying new equipment—installation requirements.

addition, all staff associated with equipment maintenance and repair must be trained appropriately by qualified representatives from the supplier.

Commissioning trials should be similar to the prepurchasing trials, covering the filling or packaging function over the full range of pack sizes and machine speeds. Again, these should be done with placebo product and packaging materials. These must be carefully reconciled and destroyed after the commissioning trials. The staff who will eventually operate the equipment when in production should carry out the commissioning trials. This will allow for staff placement and for coordination to be practiced and perfected. Quality assurance staff should monitor all aspects of the commissioning and should satisfy themselves that all GMP requirements have been met.

A report should be prepared that details the installation and commissioning work, and lists all aspects of the specification, with relevant comments and the signature of whoever checked them, including calibration certificates where relevant.

A summary of the commissioning requirements is shown in Figure 5.9.

VI. VALIDATION

Validation is a full assessment of the filling or packaging operation using product and components (which may be considered suitable for sale on completion of the validation studies). This involves a planned, closely monitored, and recorded study of the packaging operation to ensure that the necessary product quality is obtained and maintained over the full range of products and pack sizes for which the equipment is to be used.

It is not always possible to fully validate all types of products and pack sizes over a short period of time, due to either the product not being manufactured (and hence not available) and/or packaging components not being available. If the necessary validation is done as the product or packaging components become available, which may take many months, there is no problem in having a piece of equipment validated for performing only some of the functions for which it is intended.

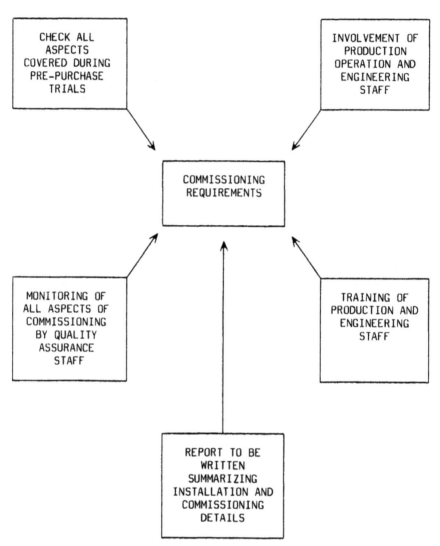

Figure 5.9. Buying new equipment—Commissioning requirements.

The validation exercise should be conducted as follows:

1. Obtain a batch of satisfactory product and relevant packaging components.

2. Set up a detailed in-process control monitoring system cov-
    ering all aspects of the packaging operation. This monitoring
    is basically a machine capability study and should be used to
    set up a realistic in-process control system. Details of the
    packaging operation machine speed, sampling points, and
    results found should be recorded. Monitor and challenge the
    functions of each of the critical devices (e.g., missing label
    detector) during the packaging operation. The whole exer-
    cise should be repeated with another batch of product and
    components (and repeated again if comparable results are not
    obtained).

3. Prepare detailed operating procedures and documents for re-
    cording the production activities. These must include recon-
    ciliation and equipment cleandown details. The cleandown
    procedure should be prepared by production, quality assur-
    ance, and engineering staff. All potential hiding places for
    rogues must be identified (including areas inside the equip-
    ment) as well as how these areas are to be accessed and
    checked. A drawing must be prepared that clearly shows all
    the areas to be checked by the production operators and
    supervisor. To ensure these are checked after each batch and
    product change on a piece of equipment, a checklist should
    be prepared requiring both an operator and supervisor's ini-
    tial for each place checked. This may seem an excessive
    amount of detail and work for a cleandown operation, but
    considering the possible consequences of just one rogue on
    the market, it cannot be underestimated. The death of a
    patient as a consequence could close a factory.

Once satisfied with the equipment operation, procedures, and a suitable in-process control system has been decided, then a validation
report can be prepared.

Once this is approved by the chief engineer, production manager,
and quality assurance manager (or suitably qualified delegates), then
the equipment is ready for production use.

A summary of the validation requirements for new equipment is
given in Figure 5.10.

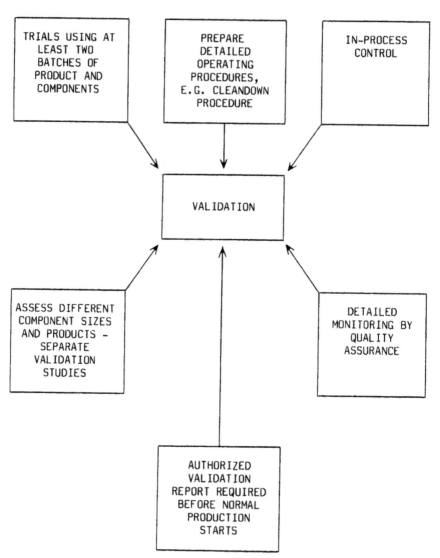

Figure 5.10. Validation of new equipment.

VII. EQUIPMENT AND COMPONENT MANUFACTURER INTERFACE

The technological advances that all industries are striving to achieve are necessary in this highly competitive world. An equipment manufacturer will try to build a packaging machine capable of a faster packaging rate. The only problem is that a high-speed machine will probably need different tolerance limits on component dimensions, usually tighter limits, or else tighter control on dimensions not considered critical on previous machines for a satisfactory packaging operation.

The packaging component manufacturer will also try to make components at a faster rate. Striving for a more efficient way of doing any job needs to be encouraged, but unless these two industries keep in step with each other, a gap may (and probably will) appear that will only cause problems for the pharmaceutical industry. Hence communication and cooperation between component and equipment manufacturers must be encouraged. This should involve extensive packaging trials for all new developments in either industry.

It has been found by experience that when the two industries work together significant quality and output benefits accrue to the pharmaceutical industry. For example, one foiling equipment manufacturer and a foil manufacturer worked closely together and the combination of the two gave very few problems, whereas with other foil and foiling equipment manufacturers, continual problems were encountered.

It should not be up to the pharmaceutical industry to bring together packaging equipment and components for the first time (to any appreciable extent), with the resulting high cost of wasted product, time, and quality problems. The only exceptions are cases of a new component designed by a pharmaceutical company. If this is necessary, then the component and equipment manufacturers should be consulted to minimize potential problems.

6

Pharmaceutical Packaging

I. INTRODUCTION

The quality control of components in the pharmaceutical premises starts at the receiving stage, which was covered in Chapter 4.

Once the components are considered acceptable by the packaging materials laboratory, the control of component quality must be maintained through each stage of handling and use, that is, from component storage and preparation to the filling, packaging, and dispatch of the product. A summary of packaging material flow in a pharmaceutical premises is shown in Figure 6.1.

Before the filling and packaging of a product can start, the planning of the work must be correct. Errors at this stage can affect the quality of the pack.

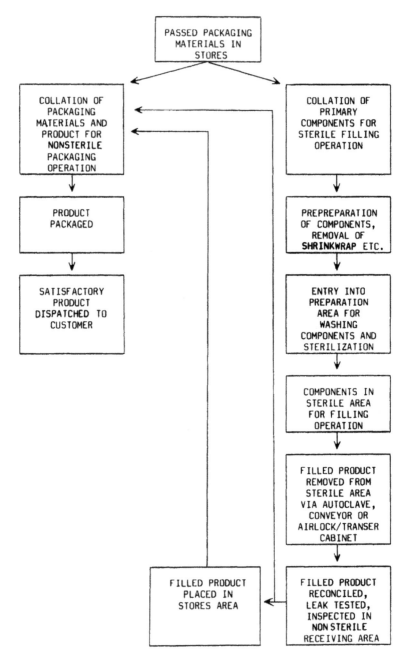

Figure 6.1. Packaging materials flow in the pharmaceutical premises.

II. PLANNING THE WORK

This is an area that is easily omitted with respect to quality control. A mistake at this stage may not be easily detected and can put a large amount of product at risk.

It is not intended that forward planning philosophy be considered, although it is worth noting that pack standardization, whenever possible, can simplify this aspect of planning.

Considering the situation in which definite orders for a product have been received, a planning officer must organize the planning of the work in conjunction with the production staff. This will help ensure the best utilization of the production equipment at the filling and packaging stages. Rationalization of orders can have both significant quality and commercial advantages, particularly if many small orders are required to be packaged. This will minimize the number of line change-overs and setup quality problems.

The planning officer must have accurate information available concerning particular customer requirements, such as special overprinting instructions. Each country will usually require everything in its own language and may use a different life for a product. An error at this stage may result in significant quantities of product being packaged before being detected, although a good in-process control system during packaging should detect a mistake like this at the set up of the equipment. Therefore, the planning officer must have a clearly defined system of ensuring that the customers special requests are available and updated as necessary.

A computer system provides the most efficient method of making this data available. These details should form part of the filling or packaging specification (covered later in this chapter) and should be given a specific reference code.

When an order is received, it must be given a specific order number to enable easy identification at the relevant stages of production. Sometimes specific order requirements do not become relevant until after the filling stage. In these instances the batch(es) of product may not be allocated to an order until the packaging is planned. The planning officer must issue the packaging instructions for each order. These must include the product batch numbers, quantities, expiry date for each batch, packaging specification reference, engineering set-up instructions, and packaging date (as agreed with the production staff). If repeat orders with identical packaging instructions are received from a customer, then a specific packaging instruction reference code could be used. This will simplify the packaging operation. The one aspect of the instructions that will vary for each order (besides overall quantities) is the expiry date for each batch of product; therefore a check to ensure these dates are correct is required. This is probably best done just prior to starting the printing or overprinting.

When the production area receives the order requirements, the work needs to be planned to ensure that everything is in the right place at the right time. This will start with the product manufacture, followed by the removal of the packaging materials from the storage area, and preparation for use in production.

III. COMPONENT STORAGE AND PREPREPARATION

It is likely, with the high level of activity within a storage area, that dust may be a problem. Therefore, all items requiring storage before use must be protected by dust covers or shrink wrapped. This protective covering should only be removed immediately prior to transferring to a production area. The covering must not be removed in a production area because of the particulate contamination risk to the production environment. Once the components arrive in the production environment, the choice of packaging for components becomes important. Opening a fiberboard container creates particulates, and with primary components such a risk is unacceptable. Therefore non-shedding packaging should be used if a lot of extra precautions and component prepreparations are to be avoided, e.g., removing the outer packaging in a prepreparation area. A set standard should be maintained in component handling at every stage, from component manufacture to use in the production area. One weak point could undo all previous precautions taken, which is a waste of time, effort, and money, in addition to being a risk to product quality.

IV. PREPARATION AREA

Components and items for entry into a sterile area should be prepared in a clean room environment. This is referred to as the preparation area.

A. Standard Required

This is an important area, requiring high cleanliness standards and carefully thought-out preparation procedures if component contamination is to be avoided.

The area should be constructed in the same manner as a sterile area for easy cleaning, that is, with a coved floor and ceiling, and with no joints or cracks in the floor or walls. It should be supplied with filtered air to class 10,000 at a positive pressure to the surrounding nonsterile areas, but negative to the sterile area. Each component-washing machine should have localized laminar flow protection to class 100 to minimize particulate contamination. The water supplied to each machine should be prepared fiber free, for instance, clean steam generated.

The staff should enter through a changing room and must wear nonshedding clothing, head covering that allows **no hair** to show, and nonshedding footwear.

Materials should be prepared prior to entry to the preparation area, and particle shedding items such as wooden pallets and fiber-board materials should not be allowed in the area. If pallets are required, they should be constructed of plastic, aluminum, or stainless steel materials only. Dust covers or shrink-wrapping covering components should be removed prior to entry to the area.

B. Preparation of Materials for Entry into a Sterile Filling Area

There are several methods of entry of materials into a sterile area. These are

1. Dry heat sterilizers (ovens and tunnels)
2. Autoclave
3. Air lock

All materials will require preparation before entry into a sterile area, whether they are irradiated components, glass or plastic primary containers, machine parts, or engineering tools.

1. Washing and collation of glass or plastic primary containers and associated primary components, then preparation for sterilization
2. Washing and preparation for sterilization of filling machine parts
3. Collation, preparation, and transfer of irradiated components into the sterile area
4. Washing and preparation for sterilization of engineering tools and spare parts
5. Preparation of lubricants for sterilization

The preparation of the components must be considered as a discrete production operation, with prime consideration to security. A component mix-up at this stage is just as serious as a filling or packaging mix-up. Therefore each component preparation operation should be conducted in a clearly defined area, with no other components in the same area. The area must be cleared of all components and checked to be clear before the next batch of components is allowed in the area. A clearly defined cleandown procedure must be followed. When using component washing machines, all possible areas harboring rogue components must be identified. A diagram showing all the areas to be checked by the operators will be necessary to ensure that critical risk areas are not missed.

Component identification must be maintained to ensure that a batch can be traced through each production stage. Therefore, each container must be identified with the component description, code reference, and batch number.

A log sheet must be used to record all aspects of the component preparation, in-process monitoring, and sterilization details. The component description, code number, batch number, and quantity must be recorded. The details of the product name, code reference, and batch number for which the components are to be used must also be recorded on the log sheet.

1. Monitoring Quality in the Preparation Area

Quality must not be ignored in this area. In-process control is necessary for all component washing and preparations.

   a. Primary Containers and Associated Components.
   (1) Vials, Ampules, and Plastic Containers. When washing these primary components, the cleanliness of the washed containers should be monitored on a regular basis (clarity checks). Also, monitor the wash and rinse water from the washing machines. It is better to identify a contamination problem at this stage than after sterilizing and filling the components with expensive product. The standards operated must be tighter than the finished filled product. This is because, no matter what precautions are taken at each stage of the production operation, there will always be a low level of particulate contamination. Also, in the filled product, particularly with vials, where the plug and overseal complete the pack, there is a potential contamination risk from the plug.
   (2) Rubber Plugs, Seals, Plastic Nozzles, and Caps. These present high risk of particulate contamination because of the electrostatic attraction of dust and difficulties in washing large quantities of small components. It is very important to monitor the particulate level in the rinse water of these components; once it is particle free, take a sample of components and perform a clarity check. Again, the standard must be tighter than in the filled product.

   b. Filling Machine Parts. All filling machine parts must be carefully washed and rinsed in particulate-free water and then closely examined to make sure that they are not damaged. Each part should have a identification number, and a checklist of all the parts required for each filling machine should be available. An error, such as wrong parts or parts missing, at this stage probably would not be detected until assembling the filling machine. This would mean a delay in filling the product. Validation of the cleaning method for product contact parts is essential when the filling machine parts are not dedicated to one product only. This is necessary to ensure that cross-contamination does not occur. A report of the validation results needs to be prepared, which will be used for preparing the cleaning procedure.

c. Irradiated Components. These will not be washed but will be transferred into the sterile area via an air lock prior to removing the outer wrapping. Each pack of irradiated components should be examined to ensure that there are no pinholes or tears in the pack that make the contents nonsterile. Also, the labelling of the outer and inner bags must be correct.

d. Engineering Tools and Spare Parts. Engineering tools will require special cleaning because of possible oil, grease, and excessive handling by engineering staff. These will require regular recleaning and sterilization. Tools are a potential particulate and bacteriological risk to the product, especially when used in close proximity to a sterile product filling head. Therefore, they should all be routinely examined for cleanliness and bacteriological contamination.

Spare parts required in a sterile area can create problems because they may be special one-off items that have not had to be prepared previously for entry into a sterile area. How they are to be cleaned and sterilized must be individually determined. Quality assurance must be involved in advising and monitoring for both particulate and bacteriological contamination of these items. Electrical or electronic items may be seriously damaged by a heat-sterilization or swabbing method, and the only alternative is to irradiate them. This would mean that certain spare parts must be irradiated and carefully stored in a sterile state if production is not to be seriously disrupted.

e. Lubricants. Lubricants will be required for all sterile-area equipment and, may be affected by a heat-sterilization method (although there are heat-resistant lubricants available). Some lubricants are self-sterilizing and, provided they are transferred into a sterile container, can be passed directly into the sterile area, although it is essential that the transfer technique is validated. Lubricants that are affected by heat or are not self-sterilizing should be sterilized by irradiation. These should be packaged into suitable nonshedding containers in the preparation area in small quantities sufficient to last 1 week in the sterile area. Each pack should be sealed in a heavy-gauge polyethylene bag that can be removed just prior to the entry into the sterile area. These packs must be clearly labelled with the type of oil used (edible, FDA approved only) and must be carefully examined in the same manner as irradiated components before entry into the sterile area.

C. Sterilization of Components, Machine Parts, and Engineering Tools

Once the preparation is complete, the components and machine parts should be placed into suitable, nonshedding, clearly labelled containers for sterilization. If stainless steel containers are used, these should have lids, but must allow both air to be removed and steam to penetrate if an autoclaving sterilization method is to be used. These containers must be suitably washed with particulate-free water and monitored.

1. Autoclaves and Dry-Heat Sterilizing Ovens

These may also be a source of particulates, therefore they should be of suitable construction to prevent such contamination. They should have an internal smooth stainless steel (not painted) finish, nonshedding door seals, and should be regularly cleaned with particulate-free materials. Also, any associated racking should have a similar finish. There should be no corners or other inaccessible places that cannot be easily cleaned.

2. Air Lock and Transfer Cabinet Entry into a Sterile Area

This is one of the highest risk areas with respect to maintaining sterility in the sterile area. A large variety of materials may enter the area via these methods, and all materials must have a clearly defined entry procedure. The main items likely to enter by this method are

1. Irradiated components and stationery items
2. Sterile product (usually powders) in sealed containers
3. Filling machines and machine parts

The attention to detail regarding the entry of the above items cannot be underestimated. A swabbing and handling procedure must be prepared for all items and must be proven to be effective by bacteriological monitoring. Following is an example of the amount of detail required for each of the above categories.

a. Irradiated Components.   These should be double wrapped (as mentioned in Chapter 4). In addition to providing extra protection during handling and transportation, the contamination risk to the sterile area is reduced if the following procedure is used:

1. The preparation area staff performing this task should wear surgical (or similar) gloves and swab their hands with a powerful bactericide before starting (ensuring sufficient contact with the bactericide to make it effective).
2. Remove the outer wrapping from the irradiated components and immediately place them in the transfer cabinet.
3. Carefully swab the outside of the pack of components with the bactericide (a swabbing tray with a layer of bactericide in the bottom is ideal), paying special attention to the heat seal (if present) on the plastic bag.
4. The swabbed pack of components should be left in the swabbing tray for a predetermined time to ensure effective sterilization, then the sterile area staff should swab the pack again before removing it from the cabinet and placing it in the sterile area.

b. Sterile Product.   The same swabbing procedures must be followed as mentioned for irradiated components, but with one further act of swabbing, when the product is in the sterile area and is about to be filled. The swabbing in the air lock or transfer cabinet may not be able to effectively penetrate all bacteriological risk areas of the seal. Therefore the operator must swab his or her hands before opening the seal, swab the rim of the container once opened, and swab the hands again before handling the opened container. The opened containers must be immediately placed on the filling machine to minimize the contamination risk. Finally, swab all parts of the seal if it is used again; if not, remove from the sterile area.

c. Filling Machine and Machine Parts.   These are probably the hardest to sterilize for entry into a sterile area without breaking sterility. The best time for the entry of large pieces of machinery is when the the whole area is to be resterilized, although preparation prior to entry will still be necessary. Preparation of the machinery will involve either using a formalizing tent or area that can be isolated and formalized. The machinery should be wiped clean wherever possible with a strong bactericide, formalized, and then all areas of the machinery monitored for bacteriological contamination. The machinery must not be moved into the sterile area until found to be bacteriologically clean. Several formalizations may be necessary.

Small pieces of machinery may be swabbed with a bactericide (using a validated technique), provided there are no electrical or electronic items attached. Electrical and electronic equipment will require formalizing using either a formalizing tent (mentioned previously) or using a plastic bag, although small items could be irradiated.

V. STERILE AREA

All product to be sold as sterile must be filled in the cleanest area possible.

A. Standard Required

The area standard required is class 100, with local laminar flow protection over the filling machines of class 100. Regular monitoring by quality assurance for both particulate and bacteriological contamination is necessary. This should include air-borne particulate and microbiological monitoring, slit sampling, settle plates, and finger-dab monitoring.

The whole area must be designed for easy cleaning, with smooth walls and a coved floor and ceiling. Fittings must be flush with the walls and ceiling wherever possible, such as light fittings with an access for maintenance from above that do not break sterility. A strict daily cleaning schedule must be followed.

A separate room for each filling machine will enable better bacteriological control. Then, if a sterility problem does occur with a product, no other products will be at risk.

Usually the biggest contamination risk to this area is from the operators and engineering staff. They must follow a strict washing and changing procedure into sterile, nonshedding, one-piece suits with complete hair covering, face masks, surgical gloves, and sterile footwear. The suits should be tight fitting around the wrists, neck, and head. All staff must be proved to be trained in the sterile-area entry procedure, including details on how to wash their hands and the method of changing into the sterile-area clothing.

Strict aseptic techniques must be followed to prevent contamination of product, components, and machine parts, such as

1. Minimal handling of components and machine parts liable to come into contact with product
2. Regular swabbing of hands, tools, and implements with a suitable bactericide
3. When a stoppage occurs requiring operator or engineering staff intrusion, all unsealed containers in the vicinity of the intrusion must be rejected
4. Engineering staff working underneath equipment must not place tools on the floor. Engineers must replace their sterile area garments after working underneath equipment.
5. All tools, implements etc. that are accidentally dropped on the floor must be immediately removed from the area, washed, and resterilized.
6. Components dropped on the floor must be reconciled and rejected.
7. All components, tools, machine parts, etc. must not be stored in a sterile area longer than a defined period (by quality assurance based on bacteriological monitoring) without resterilization.

B. Component Security

The added risk of bacteriological contamination makes component security harder to control, particularly if there is more than one filling machine in an area. A separate sterile room for each machine would be ideal, which also has bacteriological advantages, as mentioned in Section V.A. In the situation in which there is one large sterile area containing several filling machines, providing a room for each machine would require a completely new air conditioning plant layout and almost complete rebuilding of the whole area. If this is not financially viable, then an alternative method of controlling security is required. It is not advisable to separate each machine with partitions because of the cleaning problems and hence bacteriological risks. Therefore, the only alternative left is to have each machine well separated, with a clearly marked floor area. Each area is to be treated as a separate production area, where only product and packaging materials associated with the batch being filled are allowed. Both product and components must be reconciled before and after the filling of the batch of product, followed by a detailed cleandown check for rogues.

Components in the area that are waiting to be used must be clearly labelled with the component code reference, component batch number, date sterilized, product name, code reference, and batch number for which they are to be used. The date sterilized is important, because the components must not be used after the maximum period allowed before resterilization. There may be some components, such as rubber plugs, that may deteriorate if resterilized. These may have to be rejected and destroyed.

C. Sterile Filling of a Product

1. Filling Instructions

The engineering staff should have detailed instructions for the assembly and set up of the filling machine.

It is essential to have detailed filling instructions to ensure that the correct components are used with the product concerned. These should include:

1. Product name
2. Product code reference and batch number
3. Each component code reference and batch number(s) to be used
4. Product fill weight target and acceptable limits
5. Maximum time the product can be stored in the sterile area prior to filling
6. Any special filling instructions, for example, fill under ni-

trogen, carbon dioxide, etc. This should state the maximum oxygen level permitted per filled container.

When filling ampules, the maximum sealed length should be included in the instructions. This is because the ampule will have to be packaged in preformed trays and/or cartons and there will be limits to the length of ampule that can be packaged.

2. In-Process Control

Detailed in-process control instructions must be available for monitoring all critical parameters, such as

1. Appearance, including clarity checks
2. Fill weight checks
3. Headspace oxygen (when filling under an inert gas)
4. Quality of seal, i.e., appearance; with ampules, examine the seal under a polarized light to ensure that the sealing stress is acceptable. This is very important because an ampule may crack at the tip if not sealed correctly (see Figure 6.2) due to stress patterns in sealed ampules. When an unacceptable stress pattern occurs, the ampule tip may crack up to 6 months after sealing.
5. Sealed ampule length (this may only be required at the set-up stage)
6. Analysis may be required throughout filling if the product contains a suspension

All results must be recorded on an in-process control sheet. Documents must be sterilized prior to entry into a sterile area (by irradiation) and, if possible, laminated. The ideal method of recording results in a sterile area is on a computer, provided that it is designed for easy cleaning and sterilizing, i.e., it has a touch-sensitive keyboard.

In addition to the in-process checks, sealed empty containers and liquid broth trials (liquid filling only) are required for bacteriological testing.

If the complete batch of product cannot be filled in a day, then each day's fill should be identified and treated as a separate batch, especially with respect to sterility testing.

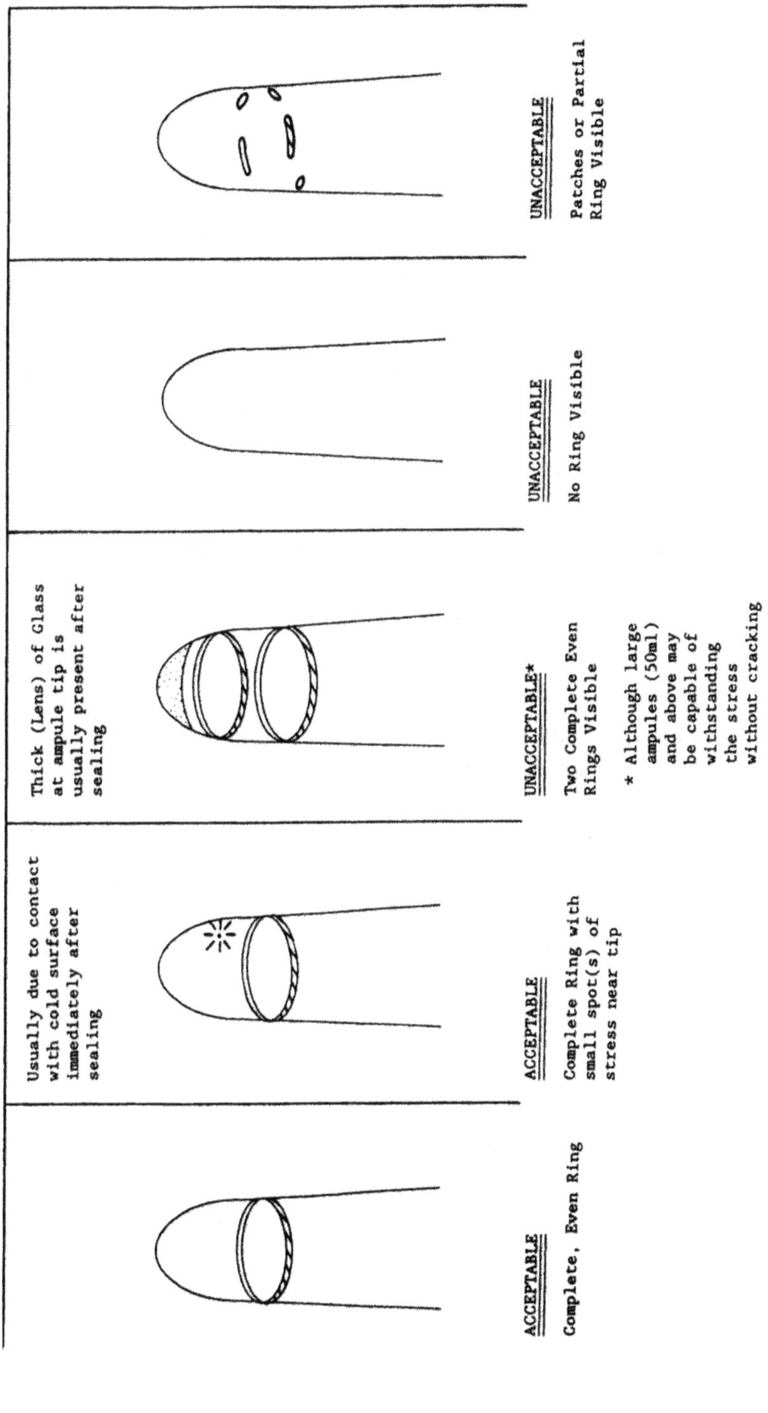

Figure 6.2. Stress patterns in sealed ampules.

3. Product Security

When filling unprinted ampules or vials, it is advisable to identify the container. Therefore, it is suggested that each container is color-ring coded (specific to each product) at the filling stage. This ring code can then be read using a code reader at the final packaging stage. This is an extra safeguard to ensure that a rogue is not packaged. If it is not intended that the filled containers be immediately labelled, then the batch or lot number needs to be printed on each container.

4. Filling Line Log

A filling line log should be available for each filling machine and should be used for recording all activities associated with each batch filled. This is required to enable the history of a batch of product to be traced. In the event of a serious complaint, this data may enable the exact cause and extent of a problem to be traced. The regulatory authorities may need to see this data, particularly if there is a product recall.

The following data should be recorded on the log sheet:

1. Product name, reference code, and batch number filled.
2. Date filled.
3. Clearance checks to ensure that all components have been cleared from the equipment area at the end of each batch or day's fill. All areas to be checked should be clearly stated.
4. Reconciliation of product and components at the end of each day's fill, as well as at the end of each batch. Reconciliation of the filled rejects must be included. Discrepancies outside the acceptable limits must be immediately investigated and recorded.
5. The batch numbers of all the components used. This is essential if a component problem is to be traced back to the supplier. This will be necessary if a customer complaint relating to a component is received.
6. Details of any machine breakdowns and action taken by both engineering and production staff. Breakdowns can be one of the major causes of product quality problems.

5. Product Removal From the Sterile Area

The filled batch of product can be removed from the sterile area in several ways:

a. Autoclave. product filled into ampules may be terminally sterilized (particularly if product degradation does not occur).

This is a severe handling stage for ampules, considering the variations in temperature and pressure to which they are subjected. Ampules that have not been sealed correctly may crack at this stage.

Each tray of filled product must be positioned in the autoclave in a predetermined and validated way. This is to ensure that each ampule reaches the required temperature for the specified length of time.

The product will be removed into the receiving area after completion of the autoclave cycle.

b. Conveyor. product that cannot be terminally sterilized (usually containing a preservative) or is filled into a container, e.g., vials or plastic bottles that cannot withstand an autoclave cycle. The product will be transferred directly from the filling machine through to the nonsterile receiving area. (The sterile area conveyor belt cannot pass through to the nonsterile area). This is the most efficient way of removing the product from the sterile area, where it can be immediately inspected (if necessary) and packaged into the final pack, ready for dispatch.

c. Via Air Lock or Transfer Cabinet. product that cannot be terminally sterilized. This may be necessary when a filling machine placement does not enable direct transfer to the nonsterile receiving area via a conveyor.

VI. FILLED STERILE PRODUCT RECEIVING AREA

Due to the variety of methods of transferring the filled product into the nonsterile receiving area, the next stage of the process can vary. The method of removal will depend on the process requirements.

This area will be used for reconciliation and possibly for crack detection, inspection, and/or packaging.

One of the main considerations must be to ensure the complete segregation of each batch of product. This requires a strict security procedure, as with all other areas, but can be complicated due to the number of batches and products that have to be removed via the same system. There will be a limited number of autoclaves and transfer cabinets available, which will be used for several batches in a day. Therefore only one batch of product should be transferred at a time and removed to a clearly defined area to await the next stage of the process.

The best method of maintaining security is when product is removed via a conveyor system.

A. Reconciliation

Product removed from the sterile area via an air-lock transfer cabinet or autoclave should be counted to ensure that the correct quantities of each batch are received, in agreement with the quantity filled. This is important if further work is required to a batch, such as inspection, because a reconciliation error found after the completion of the next stage of the process may be difficult to resolve. It may not be possible to determine whether the error occurred at the filling or inspection stage.

Product removed from the sterile area via a conveyor system may not be immediately collated. This will depend on whether the product is immediately packaged, inspected, or just placed into trays. The collation must be done after the relevant operation.

B. Crack Detection

One of the most serious defects in filled ampules is small cracks. These can be caused by poor sealing, autoclaving, or rough handling. They cannot be easily seen by visual inspection; therefore it is advisable to have a more efficient method of checking for cracks. The simplest method is to perform a leak test. This is done by placing ampules in trays lined with blotting paper. Pull a vacuum for several minutes to equilibrate the pressure, then examine the blotting paper for wet patches. Repeat the test with the ampules in an inverted position. This test cannot be used for ampules filled with a powder or freeze-dried product. The main disadvantage of this method is that there is still reliance on vigilant staff to identify the relevant leaking ampules. There are more modern techniques using sophisticated equipment such as detecting cracks using a high-voltage system. Provided the money is available to obtain such equipment, there may be significant quality and efficiency improvements. Any equipment considered must be fully validated to ensure its effectiveness and that the product quality is not affected by the technique used.

C. Inspection

It should not be necessary to 100% inspect all sterile filled product, provided an effective in-process control system is in operation; although if inspection is required, the inspection efficiency should be determined.

1. Inspection Efficiency

There are several variables to consider when determining the efficiency of an inspection operation, none of which can be overlooked if a realistic result is to be obtained. A summary of the factors that will affect inspection efficiency are shown in Figure 6.3.

Prior to starting the efficiency study, make sure that the inspectors do not know they are being monitored with manual and semiautomatic methods. The first consideration is the method of inspection; this will vary depending on the container to be inspected. For example, if inspecting a clear ampule containing a clear colorless liquid, then a manual inspection would require the following:

1. A light, either filament or fluorescent, of defined intensity and wavelength (filament is probably the best type of light).
2. A background to view against, either black or white.
3. A darkened room without distractions for the inspectors.
4. A clearly defined inspecting technique for the inspectors—a slight twist and swing to disturb heavy and light particles without causing air bubbles.
5. A set rate of inspection, that is, the time period each ampule is to be examined.

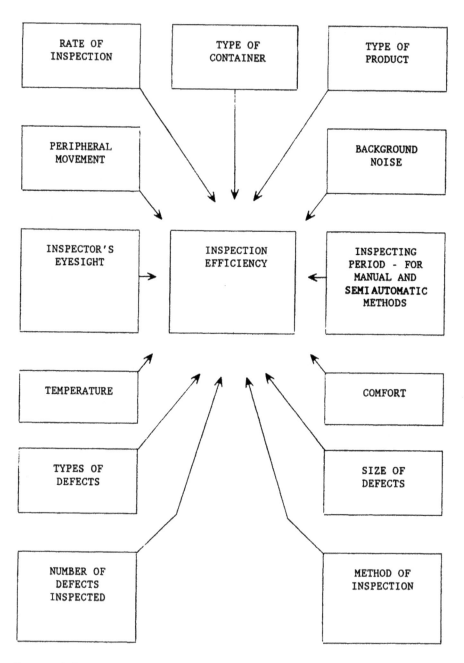

Figure 6.3 Summary of the factors that will affect the inspection efficiency.

6. Inspectors must be shown all the known types of defects, that is not only content defects such as glass, particles, fibers, and discoloration, but also ampule defects such as cracks, sealing faults (spiked tops, etc.), ring snap faults, and printing defects with printed ampules. The standard operating procedure should be: to remove all visible contaminants and container seal defects. There may be cosmetic defects such as scratched, marked, or minor print defects that can be considered acceptable. Acceptance standards should be available to the operators.

Statistical sampling of the batch of product will be required before and after inspection. The rate of inspection must not be varied during the assessment. With manual and semiautomatic inspection methods, samples will be required from the start until the end of the inspection period. This is required because the inspection efficiency will vary with time. A tired inspector will tend to have a lower inspection efficiency.

Experienced staff should examine the statistical samples for defects. If the sample size is large, at least two people should examine each sample. The defects should be carefully classified into the type of defect. With contaminants found in a product, identify and determine the size of each contaminant.

The efficiency (validation) exercise should be repeated with several batches of product, different products (particularly if colored or opaque), and different types of primary containers used.

Assessment of the results will determine the individual operator efficiency for the various products and presentations assessed at the rate of inspection monitored. If the efficiency is very low, then either a different method of inspection is required or a further assessment at a slower inspection rate. Each new inspector must be trained and be capable of inspecting to the predetermined inspection efficiency before being allowed to inspect the product.

2. Disadvantages of 100% Inspection

There are several disadvantages to relying on a 100% inspection system. These are

1. With a manual inspection method, the efficiency can never be 100% and will vary from inspector to inspector. The individual efficiency may vary a great deal, depending on the mood of the person, e.g., headache, unable to concentrate 100% of the time.

   The reliance on semi- or fully automatic inspection machines should not be overestimated. With semiautomatic machines the inspection efficiency still depends on the operator. With fully automatic inspection machines, the efficiency depends on the setting of the machine and, to my knowledge, there are none available that can satisfactorily assess both the contents for contamination and the container for cracks, marks, or print defects.

2. Product security can be very difficult to control with such a labor-intensive operation, particularly with inspection machinery. Each inspection operation would have to have the same strict security measures in operation as for a filling or packaging operation. A breakdown in the security system could lead to a rogue situation. It does not make sense to remove relatively minor defects from a batch of product and to put a critical defect into the batch in the form of a rogue.

3. With semi- and fully automatic inspection machines, there is a possibility of damage to ampules and vials during, or subsequent to, the inspection operation. These will almost certainly not be detected, possibly leading to a cracked ampule or vial on the market.

4. 100% inspection is a very labor intensive, time consuming, and expensive operation.

The alternative, an effective in-process control system, would still sometimes result in parts of a batch requiring inspection unless the suspect material is rejected. Inspecting only a small part of an occasional batch will be much easier to control than 100% inspection of every batch filled.

D. Packaging

Immediate packaging of a batch after filling is the quickest and most efficient way to operate. There is also reduced security risk from minimizing the necessary handling stages, such as the storage in an intermediate store for filled product awaiting final packaging, although it is unlikely to make the dispatch to the customer quicker because of the time required to complete the chemical and sterility testing. The requirements for packaging a product are covered in section VIII.

VII. COLLATION AREA

Before any packaging operation can take place, collation of all the packaging materials required to complete the number of packs is necessary. The area ideally should be situated between the stores and the packaging floor.

The storeperson must have clearly defined collating areas, with a strict security system in operation for each batch collation.

The storeperson must have copies of the packaging order requirements and will collate not only the required number of the various packaging items, but also the required quantity of product for packaging (although this will not be necessary when packaging a sterile product immediately after filling). All small items, which must include overprinted components, should be placed into clearly labelled containers stating the component batch number, code reference, and quantity. The various items should be placed in a large, locked pallet cage or similar container (a security seal should be used to prevent unauthorized access). The locked cage should be labelled with the customer order number, product name, and batch and code number for which the materials are required. The cage should then be transferred to the packaging floor and only be opened on the packaging line by the supervisor when packaging is to begin, and all items should be checked against the order to ensure that they are correct.

The advantages of having a separate collation area instead of trying to collate materials on the packaging line are as follows:

1. Improved security and reconciliation.
2. Minimal delays in the packaging operation, which may involve many staff. Finding that there are insufficient quantities of an item during a packaging operation will be expensive in machine downtime and in labor. The same problem found in the collation area will involve only the time of the collator.
3. Prevents confusion because excess quantities of materials are not sent to the packaging floor, leading to either expensive rejections or returns to the stores.

A record of quantities collated and returned to stores must be kept. Partial containers of components returned to stores must be clearly labelled with the exact quantity returned, item code, and batch number.

Bar coding of all packaging materials will help to improve the security at the collation stage. A bar code reader can be used to double check that the correct materials have been reconciled.

VIII. PACKAGING AREA

The packaging operation can involve many complex operations, all of which require careful control if product quality and security are to be maintained. Remember that this is the last production stage and that any errors made and not detected may not be discovered prior to dispatch to the customer.

The *packaging operation,* means either the filling and packaging of nonsterile products or the packaging stage of filled sterile products.

A. Area Standard

It is difficult to decide what standard is required for packaging a product, considering the variables of both the packaging operations and the large quantities of particulates produced during a packaging operation.

There will be no risk of contamination to a sterile filled product undergoing packaging, but with products such as tablets, creams, lotions, and oral-liquid dosage forms, where filling and packaging are performed as one continuous operation, there is a risk to the product. Therefore, depending on the type of product and packaging operation, the standard requirement of the packaging area will vary, although there are several basic standards required:

1. Each packaging line should be in a separate room (or at least separated by barriers).
2. There should be coved floors and ceilings for easy cleaning. Regular daily cleaning schedules with proven antibacterial cleaning agents are required.
3. The packaging operators should wear nonfiber-shedding overalls that have a tight fit around the neck and sleeves. There should be no external pockets above waist height. This will reduce the risk of pens, etc. being packaged with the product. Complete hair covering should be worn at all times. Jewelry, such as bracelets and earrings, should not be worn, and personal medication, such as tablets, must not be taken into the packaging area.
4. The filling part of the packaging operation should be enclosed and supplied with filtered air to the standard considered necessary for the product concerned.
5. Basic precautions prior to filling can be operated, such as blowing the container with filtered air immediately before filling (ideally in the inverted position). Washing of the containers should not be necessary, provided that the correct production and operation standards are in use in the supplier's premises and during subsequent transport and storage.

B. Packaging Instructions

It is likely that there may be many variations in the packaging items required for each product, depending on the market and pack size. Therefore clearly defined packaging instructions must be available to the packaging line supervisor and engineering staff who are setting up the line. The line supervisor will require the packaging specification, specific batch details, and the customer requirements. The engineering staff will require the line set-up instructions.

1. Packaging Specifications

These are the detailed requirements for completing the pack. Each specification must have a specific reference code. This will enable the planning staff to quote the specification reference when scheduling the customers orders.

The following details must be included in the specifications:

1. The product name, strength, and reference code to be packaged.
2. The pack size and number of product items to be included in each pack, that is, the number of tablets, ampules, vials, etc.
3. A description and reference code of each packaging item to be included in the pack. Include the specific overprinting instructions (when required), i.e., print details, size, color, and position.
4. Special precautions to be taken during the packaging operation. For example, the packaging of a moisture-sensitive tablet may need to be performed under low-humidity conditions. State the maximum humidity allowed.
5. Detail the in-process control system to be operated. This will vary depending on the complexity of the packaging operation. The example mentioned in point 4 will require monitoring of the low-humidity environment and of the moisture content of the tablets.

2. Specific Batch Details

The line supervisor will need to know the batch numbers and expiry dates for each batch to be packaged. This is to ensure that the correct batches are packaged for a specific customer, with the correct expiry date printed on the packs. The life to expiry may vary depending on the market, climatic conditions, and recommended storage temperature.

3. Customer Requirements

The customer order may be to package all the product in each of the batches allocated or in an exact number of packs. The quantities of components required for each batch to be packaged need to be stated.

This quantity must include overages based on expected wastage due to line set up, breakdowns, and in-process checks.

4. Engineering Set-Up Instructions

The engineering staff should have clear instructions on how to set up each piece of equipment in the packaging line. This is essential when the same line is used for several different pack sizes. The instructions should have the same reference code as the packaging specification.

The set-up instructions should state each machine part required by name and reference number; therefore, each machine part will need a specific number. If not numbered by the manufacturer, then a number will either need to be stamped on (or affixed in some other suitable way) to ensure that the part identity is not lost. The machine parts will need to be kept in a secure (locked) area when not in use. Ideally, each set of parts should be placed in one container with the relevant packaging reference number(s). Each part should be carefully visually examined prior to fitting to the machine. This will ensure that time is not wasted in setting the machine up to work correctly, which will probably cause packaging quality problems, as well as lost production time. It is worth remembering that the machine will only work as well as the engineering staff can set it up. Therefore, the set-up instructions should include the exact adjustments required for each part fitted. Ideally diagrams showing the exact placement of each part should simplify the instructions and should be particularly useful for training new engineering staff to set up the equipment.

When the packaging is complete, the machine parts must be cleaned after the removal from the machine. The parts should be carefully examined before they are put into the secure storage area. This will enable damaged parts to be replaced before they are required again. Discovering damaged parts just prior to their being required will mean a delay in setting up the packaging equipment.

C. Product Quality and Security

Once the packaging line is set up and the correct packaging instructions are available, product quality and security must be maintained throughout the packaging stage. The previously mentioned systems for line cleandown, in-process control, reconciliation, and restrictions on staff movements must be implemented and proved to be effective. The critical devices need to be identified and proved to work correctly, and further security measures can be implemented, such as bar coding of the printed packaging items, missing print/component detectors, and automatic weight checking. Consider each of these areas in detail.

1. Critical Devices

A critical device is any device that, unless it is working correctly, could affect product quality. Each device must be identified and either calibrated or challenged on a regular basis to ensure that it is working to within the specified limits. The frequency of calibration or challenging is decided by the device reliability. The requirements concerning calibration were covered in Chapter 5, Section IV. The challenging of devices has not yet been mentioned; this should be performed by the packaging staff. A missing print detector would be challenged at the beginning of each day. This would involve passing a set number of packs through the detector (probably only about ten), one of which would have the print missing. This pack must be automatically rejected whenever presented to the detector at the maximum packaging speed. In the event of a challenge failure, an instrument technician must be called and production must not continue until the critical device is repaired, unless a suitable alternative method of ensuring product quality is maintained, possibly by 100% on-line inspection or by replacing the device. The instrument technician needs a standard to work to when adjusting the sensitivity of a device, since it is not generally a viable proposition to have missing print detectors with the ability to read the print. What is actually happening is that they are detecting a black mark (if black ink is used) on a label or carton. Therefore, the detector sensitivity will be set to read a certain size and intensity in a particular position. The technician must have a standard card with a print mark of the required intensity to check the device sensitivity. (It is always worthwhile considering whether a device is really doing a useful function or whether it is better to remove it. With such detectors, it is inevitable that some packaging items with faint or smudged print that cannot be interpreted will end up on the market).

2. Bar Coding

All printed packaging items (e.g., printed containers, labels, leaflets, and cartons) should be bar coded, with the code reader set to read each item immediately prior to including into the pack. With ampules and vials, a ring-coding system or a similar method should be used. The coding should take place at the filling stage to ensure the correct ring coding for the product (see Section VC3). The ring code should be read immediately prior to insertion of the product in the pack. Code readers will serve three main purposes: ensuring the item is correct, ensuring the item is included in the pack, and also counting the number of items used. A bar code reader will be classed as a critical device and must be challenged regularly (see the Section VIIICI).

3. Missing Print and Missing Component Detectors

If these detectors have been proven to work correctly, they can give additional assurance of a satisfactory pack on the market.

A missing print detector will help to ensure that each component overprinted with batch number and expiry date has been printed. An unprinted component must be automatically rejected from the line.

The missing component detectors will serve several useful purposes:

1. A missing leaflet detector will ensure that a leaflet is inserted in each pack, provided it is situated to detect the leaflet insertion and not the presentation.
2. A missing container detector will ensure that the product has been inserted into the carton.
3. A missing carton detector will ensure that a filled container is not presented when a carton is not available, avoiding possible breakages or filled-container damage.
4. A missing label detector will ensure that each container has a label attached.

With the missing component, leaflet, label, and pack detectors, an automatic reject system must operate to remove the defective packs from the line.

A complete pack automatic checkweighing system on the end of a packaging line can be incorporated as a final check for missing items from a pack. This will also detect high and low fill weights.

D. Packaging Equipment Log

Each packaging line must have an equipment log to record all activities carried out during each packaging operation. The details required will be

1. Product name and batch number
2. Productive time and date
3. Record of all maintenance, calibration, and challenges, including comments concerning any problems and actions taken
4. Details of all line packaging problems and equipment breakdowns, including actions taken
5. Each entry in the log should be signed by whomever made the entry

These records should be kept along with the batch manufacturing and packaging documents for well beyond the life of the product. It is essential data for investigating market complaints, especially when problems and breakdowns occur. These are the times most likely to result in defective product or packs being produced, and if not dealt with correctly, will lead to complaints.

Consideration should be given to having a computerized packaging-equipment log system. This would enable the rapid assessment of productive time, and identify specific problems, such as the continual breakdown of a piece of equipment. Troublesome equipment should be replaced or modified.

E. Packaging Sheet

It is unlikely that all details of the packaging operation can be included in the equipment log, but if they can, this is ideal for minimizing the numerous documents required for each batch of a pharmaceutical product. A packaging sheet is necessary to record the following:

1. Line clearance checks at the start and end of each batch packaged.
2. Reconciliation of the product and each of the components at the start and end of each batch.
3. In-process control results from production and quality assurance staff, e.g., print checks, tablet counts, weight checks, etc.
4. Description, reference code, and batch numbers of all printed and primary packaging components used. This is essential for tracing a batch of components used back to the supplier in the event of a customer complaint.

All data entered on the packaging sheet must be signed and the completed sheet checked and signed by the line supervisor and the person responsible for the production area. Just because a document is completed correctly does not mean the work has been done correctly; therefore, well-trained staff and supervisors are required. Also, all critical functions should be double checked (by the supervisor), such as line clearance checks, as mentioned in Section D.

F. Keeping Samples

Samples of the complete pack must be retained beyond the life of the product under controlled conditions of temperature and humidity. Samples should be stored at the expected market storage temperatures. When a product is supposed to be stored in a refrigerator, samples should also be stored at room temperature. This will enable customer complaints relating to product degradation to be assessed. Keeping samples are retained for investigating customer complaints and for the periodic check on product stability. Sufficient packs should be available to enable a full assessment of the product.

G. Casing and Dispatch of the
Finished Product

Once the packaged batch has been completed and reconciled, the individual batch identity should still be maintained by keeping each batch on a separate pallet. This is because it is important to know the destination of every pack in a batch. In the event of a recall, it is essential to know who to contact. Part of a batch sent to the wrong destination will be difficult to trace. A record of the exact quantities of every batch of a product sent to a customer must be retained with the batch documents.

1. Casing

The product must be suitably cased to withstand the transport method used. Product damaged in transit may be a risk to the patient, in that a cracked vial or ampule may be used without detection.

The casing staff must have the following details:

1. Customer name and address
2. Customer order number
3. Product or products to be dispatched, including the total quantities and batch numbers
4. Casing instructions, including the size and type of containers to be used and labelling instructions

Each batch of product should be kept together. One product should be cased at a time in a clearly defined area, to ensure that the wrong product or rogue packs are not sent out to the customer.

The casing labels should be clearly typed and checked. A simple computer program could be used for printing these. If the cases have the labelling details stencilled on them, then these should be checked. Stencilling is an ideal method of labelling to ensure that the details are not accidentally removed during transportation. The labelling must include the product name, quantity of packs, customer name, and address. In addition, special transportation instructions are required, such as *Store this way up, Store in a cool, dry place, Store below 5°C but do not deep freeze*, etc.

To ensure that the correct instructions are carried out for each casing operation, it is suggested that the packaging instruction stay with each batch of product from the component and product collation through to casing and dispatch. Alternatively, the details could be available on a computer and controlled for each customer order number from the planning section (see Section II).

2. Dispatch

The type of transport used should be of a high standard. Cleanliness and limiting exposure of the product to the elements, e.g., rain or extremes of temperature, are important aspects to consider.

It is suggested that trucking containers dedicated for transporting pharmaceutical products, or at least similar materials, be used. Dirty containers with a strong, unpleasant odor may affect the product.

IX. STAFF TRAINING

The staff training requirements are basically the same as for the packaging suppliers (Chapter 3, Section IIA). In addition, there is one particular aspect of training to be covered, staff working in a sterile area.

A. Sterile Area Training

Prior to entering a sterile area, all staff need to be familiarized with the following:

1. What a sterile area is, how it is obtained, and maintained
2. Reason for having a sterile area, covering the potential risks to the products and hence the customers
3. Main sources of bacteriological and particulate contamination, that is, everything entering the area and, in particular, people
4. Methods used to minimize the contamination risks and monitoring systems used

This awareness concerning sterile areas is essential if staff are to be committed to strictly adhering to the sterile area procedures.

The next part of the training is the changing procedure for entry into the sterile area. This should be shown and practiced in a nonsterile environment. Ensure that personal hygiene is included. For example, washing hands will be an essential part of the changing procedure. State and show exactly how hands must be cleaned.

When production staff are working in a sterile area, they must be trained in both component- and product-handling techniques to minimize particulate and bacteriological contamination. When intrusions are required into a filling machine while filling product, reject the unsealed containers in the region of the intrusion.

Special training for engineering staff is required, such as swabbing of tools before adjusting a machine and not placing tools on the floor.

This attention to training for all activities in a sterile area is essential if serious bacteriological or particulate problems are to be avoided.

B. Training Records

Training records must be kept for all staff, detailing both on- and off-the-job training. With on-the-job training, each specific activity needs to be recorded, for example, a packaging line operator may be trained to operate a machine and to perform in-process control checks. Therefore, each of these job functions must be recorded, stating when the training was started, completed, and who carried out the training. An example of off-the-job training could be GMP training.

A matrix showing all the staff in an area and in what job functions they are trained would be useful to the supervisor for determining quickly who is suitable for a particular production operation and deciding the team of staff required. A simple computer system could be used, and this could be easily updated when staff movements occur.

7

Customer Complaints

I. INTRODUCTION

The main consideration when producing pharmaceutical products is the safety and welfare of the patients using the products. Therefore, the feedback obtained from pharmacists, doctors, and patients is very important, especially if they are complaining. It is essential that customer complaints are investigated and acted upon immediately, even though many of them may be minor. Until investigated, the importance of the complaint may not be fully appreciated.

II. COMPLAINT PROCEDURE

A complaint procedure must be available, detailing how to process a complaint and the data required. Figure 7.1 shows the processing of

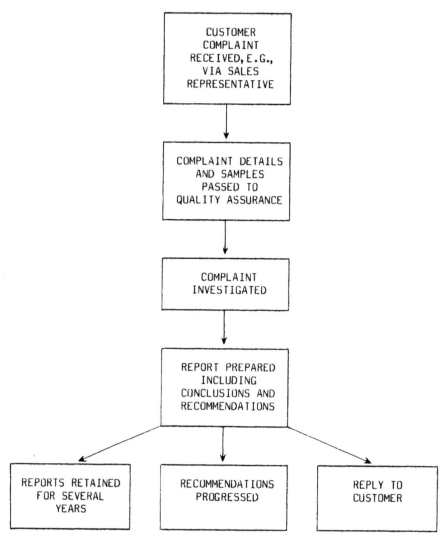

Figure 7.1.   The progression of complaints.

complaints. It is essential that all staff who are liable to be presented with a complaint, e.g., sales representatives, know what procedure to follow. All complaints should be investigated by a senior quality assurance officer who is fully trained in how to ensure that the maximum information is obtained from sometimes very limited data.

A. Complainant Communication

The following data are required from the complainant before the investigation can start:

1. Name and address of the complainant.
2. Product name and batch number.
3. Nature of the complaint. This will enable a preliminary assessment of the seriousness of the complaint, i.e., whether it is a potential recall. If it is a possible recall situation, then the investigation must begin immediately.
4. Always ask for samples to be returned. Not having samples necessitates relying on testing or examining keeping samples. Many complaints cannot be satisfactorily answered without returned samples, particularly if the batch number is not known.

B. Complaint Investigation

1. The complaint investigator should first determine the quantities and carefully check the appearance of the samples (including microscopic examination where relevant). This can be very important, and staff presented with complaints (other than the investigator) must not open or touch the samples. Important data may be lost if tampered with by inexperienced staff. For example, supposing a complainant stated that the tablets received had varying color on the film coat. The cause may be easily identified by obtaining an unopened pack. If the tablets were found to be faded down one side of the container, this could be due to exposure to direct sunlight. This would ensure a rapid answer and hence solution to the complaint. An inexperienced person tipping the tablets out of the container and then back in again before they reach the complaint officer would have destroyed this valuable data.

2.   If analysis of the complaint samples is required, then always test with the keeping samples and a separate control batch. This will eliminate any doubts about the accuracy of the results, particularly when unexpected results are obtained for the complaint sample.

3.   When a component manufacturing problem is suspected, the supplier must be informed of the complaint details, component batch number, and samples to enable an accurate assessment of the fault, and hence the corrective measures required. Notification of the complaint should be through the problem notification system mentioned in Chapter 3, Section IV. The supplier must be able to determine the extent of the problem, and hence the batch numbers and quantities supplied to the pharmaceutical company. This data is essential if a recall is considered necessary. For instance, if a rogue label or carton complaint was received and considered to be the fault of the supplier, the pharmaceutical company could then determine the batches of product to recall from the market. If the in-process control recording system described in Chapter 3, Section II B. is used in the supplier's premises, then this data should be available.

4.   A full assessment of the batch documents should be carried out with all complaints. If the complaint was a rogue, incorrect print, potency, or other such serious complaint, then the product packaged at the same time must be investigated to determine the extent of the problem.

5.   Cross-check with previous complaints to see if there is a particular pattern, e.g., repeated complaints of adverse reactions with one batch or product.

6.   The relevant production, engineering, and quality assurance staff must be involved in the complaint assessment by the quality assurance complaints officer, who will collate all the data together to produce a full report, including recommendations on any improvements required to prevent further complaints. A copy of this report must be sent to all relevant senior staff in the factory.

Figure 7.2 summarizes the data required when investigating customer complaints.

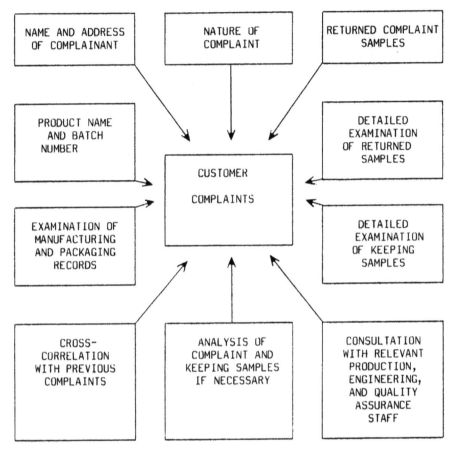

Figure 7.2. Data required when investigating customer complaints.

A record of all complaints received, including the reports, should be kept for several years. A computerized system would be ideal for rapid recall of previous complaints, or for identifying complaint patterns.

7. A summary report stating the conclusions (including the justifications leading to the conclusions) and actions to be taken to prevent recurrence must be sent to the customer.

8. A follow-up system to ensure all actions regarding complaints are carried out, whether component modifications, product manufacturing modifications, or packaging changes.

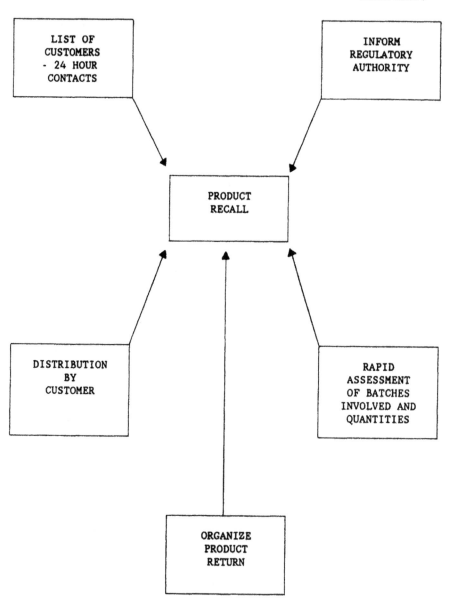

Figure 7.3. Actions required when recalling a product from the market.

For instance, suppose there are repeated complaints of labels falling off containers, this would result in the necessity to change the adhesive used on the labels. A simple form could be used for initiating the change. This would be completed by quality assurance and forwarded to the packaging development section for progressing.

C. Product Recall

A clearly defined procedure for the recall of a product must be available. The quality assurance manager is responsible for coordinating the recall and hence must be fully aware of the whole recall mechanism. This procedure must clearly define how the following actions can be completed

1. How to obtain details of where each batch or partial batch has been sent, including exact quantities and when dispatched (this must be possible at any time, day or night)
2. Name and address of each customer and a 24-hour contact
3. A rapid assessment of all batches involved
4. Regulatory authority contact—immediate notification is necessary
5. Distribution of the product by the customer
6. How the product is to be returned to the pharmaceutical company
7. When the product is returned to the pharmaceutical premises, it must be reconciled and quarantined

Figure 7.3 summarizes the actions required when a product recall is necessary.

Index

Accelerated stability trials, 9
Acceptable quality levels (AQLs)
  appearance standards, 25, 26
  dimensional standards, 38, 39
Ampule seal quality, 146, 147
Approved suppliers, 57
Artwork
  preparation rules, 21-24
  security in suppliers premises,
    75
Auditing supplier premises, 45-
    60
  auditing sequence, 49
  auditor requirements, 45-48
  auditor problems, 57, 58

[Auditing supplier premises]
  GMP requirements (see Good
      Manufacturing Practice)

Bar coding, 161
Batch definition, 24

Calibration (see Maintenance and
    calibration)
Carton, drawing of, 22
Casing and dispatch of packaged
    product, 163-165
Chemical testing of components,
    40, 41

Code of Federal Regulations (CFR) (*see* Regulatory requirements)

Collation area, 155, 156

Commissioning (*see* Equipment commissioning)

Compatibility and customer usability, 39, 40

Compatibility, product/pack, 8, 9

Complaints, customer, 167-173
  investigation of, 169-171
  procedure, 167-173
  product recall, 172, 173
  progression of, 168

Component
  primary, definition of, 24
  secondary, definition of, 24

Component artwork (*see* Artwork)

Component design (*see* Packaging design)

Component drawing, 19, 21
  examples of, 21, 22
  preparation rules, 19, 20

Component problem investigation, 95, 97-100
  action summary, 99
  investigator requirements, 95, 97
  investigation approach, 97, 98
  supplier notification, 81-83
  notification form, 82

Component/product validation, 8-12

Component quality control (*see* Quality control)

Component specification
  agreement with supplier, 81
  preparation of, 19-43

Component standardization, 3

Component sterilization (*see* Sterilization)

Computers
  equipment control, 114
  measuring equipment interface with, 39
  production planning with, 135

Crack detection in ampules, 150, 157

Critical devices, (*see also* Maintenance and calibration)
  challenging of, 160
  definition of, 160
  identification of, 110, 111

Critical dimensions (*see* Dimensions)

Customer requirements
  pack design, 12, 13
  pack usability, 39, 40
  packaged product, 158, 159

Customer complaints (*see* Complaints)

Dimensions, 26-39
  computerized measuring equipment, 39
  critical and noncritical, definition of, 26
  measurement standards, 37-39
  measuring component, 28-35
  measuring equipment, 28, 29

[Dimensions]
  precision and accuracy, 35-37
  measuring techniques, 29-35
Documentation
  equipment validation, 129
  method layout, 94, 95
  testing methods, 94
Drug Master File (DMF) (see
    Regulatory requirements)

Equipment capability assessment,
    68-70
Equipment choice, importance of,
    107, 108
Equipment commissioning, 125,
    127
  requirement summary, 128
Equipment/component manufac-
    turer interface, 131
Equipment installation, 120-125
  preinstallation considerations,
    120-122
  requirement summary, 126
Equipment purchase, 116-120
  predelivery inspection, 120
  prepurchase trials, 118-120
    requirement summary, 121
  supplier choice, 116-118
Equipment specification, 108-116
  preparation of, 112-116
  production requirements, 108-112
  requirement summary, 117
Equipment validation, 127-130
  requirement summary, 130

Filling (see Sterile product filling)
Filling machine parts
  preparation for sterilization,
    139
  sterilization of, 142, 143
Formulation details for compo-
    nents, 7, 8

Gamma irradiation (see Irradiation
    sterilization of compo-
    nents)
Gang printing, 76
Glass vial
  chemical testing, 40
  flange measurement, 29, 30
  flange profile, 30
  measurement of, 29-33
  seal integrity, 9, 10
  sterile seal
    components required, 26
Good housekeeping in supplier's
    premises, 77, 78
Good Manufacturing Practice
    (GMP)
  component supplier deficien-
    cies, 62-81
  component supplier require-
    ments, 50-57
  equipment design requirements,
    113, 114, 116

In-process control
  advantages, 68, 69
  component preparation stage,
    139-141

[In-process control]
 definition of, 63
 during component manufacture, 63-73
 equipment validation stage, 129
 packaging stage, 160
 sterile filling stage, 146-147
Inspection of filled product, 151-154
 disadvantages, 153, 154
 inspection efficiency, 151-153
 factors affecting summary, 152
Irradiation sterilization of components, 100-105
 dosimeters, use of, 102, 103
 irradiation indicator, 101, 102
 irradiation validation, 102, 103
 preparation for, 101-103
 production use of, 105
 receipt after irradiation, 103, 105
 sampling, 103, 105

Keeping samples, 163

Laboratory for testing components, 85-105
 component testing, 91-93
 layout, 87-90
  example of, 89
 location, 86, 87
  example of, 87

[Laboratory for testing components]
 recording results, 92, 93
 sentencing, 93, 94
 workbook example, 95
Legibility (*see* Print requirements)
Light transmission, 11, 12
Lubricants
 choice of, 114
 sterilization of, 140, 141

Machine clearance, 74
Machine parts (*see* Filling machine parts)
Maintenance and calibration, 123, 125
 planned preventative, 123, 125
 quality related, 123
Marketing
 considerations during packaging design, 2
 involvement with packaging design, 4
Material of construction
 component design considerations, 7, 8
 component manufacturing considerations, 76, 77
Missing print and component detectors, 161, 162

New Drug Application (NDA) (*see* Regulatory requirements)
Noncritical devices, 115
Nonsterile product/pack validation, 10-12

Nonmoulded components measurement standards, 38, 39

Optical projector
measuring components with, 28-30
precision and accuracy, 35-37

Packaging
immediately after filling, 155
Packaging area, 156-165
standard required, 156, 157
Packaging component (*see* Component)
Packaging design, 1-19
design coordinator responsibilities, 1, 2
Packaging equipment, 107-131
Packaging equipment log, 162
Packaging instructions, 157-159
batch details, 158
engineering set-up instructions, 159
specifications, 158
Packaging validation trials, 5, 7
validation protocol, 5
Personnel training (*see* Training)
Planned preventative maintenance (*see* Maintenance and calibration)
Planning
pharmaceutical packaging, 135, 136
Plastic primary components
chemical testing, 40, 41

Polymer mix preparation, 76, 77
Precision and accuracy
optical projector, 35-37
table of results, 36
Preparation area for sterile area
entry, 137-143
monitoring quality, 139-141
preparation of materials, 137-141
standard required, 137
sterilization of materials, 141-143
Primary component definition, 24
Printed components (*see also* Artwork)
requirements, 13, 14
Printing plates
preparation of, 75, 76
security of, 76
Problem investigation (*see* Component problem investigation)
Product considerations during component design, 12
tablet shape, 12
Product quality
filling stage, 146-148
packaging stage, 159-163
Product recall (*see* Customer complaints)
Product security during filling, 148
Product stability
considerations during packaging design, 8-12
nonsterile product, 10-12

[Product stability]
  sterile product, 8-10
Purchasing policy for components,
  60

Quality audit (*see* Auditing supplier premises)
Quality control
  component suppliers premises,
    61-83
    contact person, 61
    GMP deficiencies, 62-81
  component testing and standards,
    24-41
    appearance standards, 25,
      26
    chemical testing, 40, 41
    dimensions, 26-39
Quality control laboratory (*see*
  Laboratory for testing
  components)
Quality problem investigation
  (*see* Component problem
  investigation)
Quality related maintenance (*see*
  Maintenance and calibration)
Quarantine area
  supplier component storage,
    51
  manufactured components, 55

Raw material storage
  supplier GMP requirements,
    50, 51
  summary of, 52

Recall (*see* Complaints)
Reconciliation
  component manufacture, 56,
    73, 74
  container labelling, 55
  filled product, 150
Recording results
  component testing, 92, 93
    workbook front cover, 95
  in-process control during com-
    ponent manufacture, 70
  in-process control check
    sheet, 71
  in-process control during pro-
    duct filling, 146
  packaging equipment log, 162
Reference standards
  for calibrating equipment, 123
  standard weights and gauging
    blocks, 89, 90
Regulatory requirements, 14-19
  annual NDA reports, 15, 17
  Drug Master File (DMF), 15,
    17, 18
  New Drug Application (NDA),
    14-16
  supplemental NDA, 14-17
Reports
  audit, 57
  complaint, 170
  equipment installation and com-
    missioning, 127
  equipment validation, 129
Rogue components
  equipment design considera-
    tions, 113
  minimizing risk during com-

[Rogue components]
ponent manufacture,
73-75
minimizing risk during product
filling, 148

Safety
customer, 12, 13
equipment maintenance, 123
Sampling
area in laboratory, 88
components, in-process, 70-72
components, batch, 91
components, irradiated, 103,
105
Seal integrity, 9-11
Secondary components, definition
of, 24
Security
component manufacture, 73-
77
component preparation for filling, 138
component testing, 91-93
component, during sterile
filling, 144, 145
filled product inspection, 154
filled product packaging, 159-
162
Sentencing of components, 93,
94
Service manuals for equipment,
115, 116
Services for equipment, 113
Spare parts
for equipment, 125
preparation for sterilization,
139, 140

[Spare parts]
sterilization of, 142, 143
Specifications
component, 19-43, 81
preparation of, 19-43
layout of, 41-43
equipment, 108-116
filling product, 145, 146
packaging, 157, 158
Stability testing
accelerated, 9
keeping samples, 163
product/pack, 8-12
Staff training (see Training)
Standardization
components, 3
advantages, 3
equipment, 115
Standard operating procedure,
63, 129
Standards
component testing, 24-41
appearance, 25, 26
measurement, 37-39
packaging area, 156, 157
preparation area, 137
sterile area, 143, 144
suppliers manufacturing premises, 80, 81
Statistical sampling (see Sampling)
Sterile area
air lock entry to, 141-143
component security, 144,
145
filling instructions, 145,
146
filling of product, 145-149

[Sterile area]
    in-process control, 139-141
    material, entry to, 137-143
    product removal from, 149
    staff entry, training, 165, 166
    standard required, 137
Sterile filling of product, 145-149
Sterile product
    pack validation, 8-10
    sterile area entry, 142
Sterilization
    autoclave, 141
    dry heat sterilizers (ovens), 141
    irradiation of components, 100-105
    swabbing, 141-143
    terminal sterilization, 149
Storage area
    components awaiting testing, 88
    components in pharmaceutical premises, 136
    supplier raw materials - GMP requirements, 50-52
Stress in ampules, 146, 147
Supplemental NDA (see Regulatory requirements)
Supplier contact person, 61
Suppliers of components
    purchasing policy, 60
    quality auditing of, 45-60
    quality control at, 61-83
    requirement summary, 59, 60
    working relationship, 60
Swabbing of materials, 141-143

Tablet foiling and packaging, 5-7
    stages of, 6
Tablet shape, packaging implications, 12
Tamper evident pack, 2
Testing components (see Quality control)
Testing methods (see Documentation)
Tooling control for component moulding, 78-80
    commissioning of, 79, 80
    manufacturer requirements, 78, 79
    refurbishing checks, 79, 80
    wear of, 80
Training
    auditor, 46, 47
    equipment use, 125, 127
    measuring components, 28-37
    sterile area entry, 165, 166
    supplier staff, 62-67
        engineers, 63, 65, 66
        production operators, 63, 64
        production supervisors, 66, 67
        quality assurance, 63, 65
Training officer, 62
Training records, 63, 166

United States Pharmacopeia (USP), 10

Validation trials
  component/product, 8-12
  equipment, 127-130
  irradiation of components,
      102, 103
  packaging, 5-7
    protocol for, 5
Vial measurement, 28-35

Vendors (*see* Suppliers of compo-
      nents)

Warehousing GMP (*see* Storage
      area)
Water vapor permeability, 10,
      11